AF826152

Taking Technology to the Market

Taking Technology to the Market

A Guide to the Critical Success Factors in Marketing Technology

IAN LINTON

Routledge
Taylor & Francis Group
LONDON AND NEW YORK

First published 2012 by Gower Publishing

Published 2016 by Routledge
2 Park Square, Milton Park, Abingdon, Oxon OX14 4RN
711 Third Avenue, New York, NY 10017, USA

Routledge is an imprint of the Taylor & Francis Group, an informa business

Copyright © Ian Linton 2012

Ian Linton has asserted his moral right under the Copyright, Designs and Patents Act, 1988, to be identified as the author of this work

All rights reserved. No part of this book may be reprinted or reproduced or utilised in any form or by any electronic, mechanical, or other means, now known or hereafter invented, including photocopying and recording, or in any information storage or retrieval system, without permission in writing from the publishers.

Notice:
Product or corporate names may be trademarks or registered trademarks, and are used only for identification and explanation without intent to infringe.

British Library Cataloguing in Publication Data
Linton, Ian.
Taking technology to the market : a guide to the critical
success factors in marketing technology.
1. New products--Marketing. 2. High technology
industries--Marketing.
I. Title
620'.00688-dc23

Library of Congress Cataloging-in-Publication Data
Linton, Ian.
Taking technology to the market : a guide to the critical success factors
in marketing technology / by Ian Linton.
p. cm.
Includes bibliographical references and index.
ISBN 978-1-4094-3595-2 (hardback) -- ISBN 978-1-4094-3596-9 (ebook)
1. High technology--Marketing. 2. High technology
industries--Marketing. 3. New products--Marketing. I. Title.
HC79.H53L56 2012
658.8--dc23

2012011065

ISBN 13: 978-1-4094-3595-2 (hbk)

Contents

About the Author

Ian Linton has been working with technology companies as a professional writer for more than 20 years. He has been involved in marketing communications programmes for global technology clients including AT&T, IBM, ICL, Cisco, HP, Dell, NEC, McAfee, HP, Worldpay, Siemens and Mercury.

His experience covers call centre technology, convergence, data centre solutions, cloud computing, networking, knowledge systems, real time working, telecommunications, mobile communications, IT security, IT services, IT consultancy, collaboration, conferencing, unified communications and machine-to-machine communications.

Introduction

The technology sector is a global business characterised by short product cycles, rapid change, longer-term customer relationships, complex decision-making processes, high levels of collaboration and partnership with customers and the supply chain, diverse channels to market and an emphasis on the value of information. Those factors make the marketing of technology products and services a distinct discipline within the overall marketing spectrum.

With intensifying competitive activity and continuing budget constraints, technology marketing teams are under pressure to be more accountable and deliver measurable results that demonstrate an effective return on investment. To add to the complexity, the market for technology products and services is global, with continuing growth in both developed and developing territories.

When a company takes a technology product to market, it must coordinate a number of different functions including sales, marketing, pricing and market research to ensure a successful launch. The marketing team must ensure that it is fully prepared to sell, deliver and support the product in line with customer needs. That requires a plan which should deliver effective marketing communications, a highly trained and motivated sales force and an effective support infrastructure in place. An effective marketing plan will result in faster time to market, revenue and profit.

Successful technology marketing builds on an understanding of the customer needs and pain points that drive market demand. It uses targeted highly personalised product communications to demonstrate how the product meets those needs. The marketing plan focuses on the product features and benefits that differentiate the product from competitors and represents best value for customers and prospects. It utilises professional services and shared knowledge to drive greater value and increase customer satisfaction.

Taking Technology to Market provides a practical guide to the critical success factors in marketing technology. It utilises a project-based approach, providing comprehensive guidelines for key strategic and tactical marketing programmes. The book will help you improve their chances of developing a winning marketing programme by providing essential steps to success and insight into best practice.

Individual chapters provide self-contained guides to planning specific marketing tasks. The range of tasks covers the most common challenges facing marketing teams in technology companies. The book will help you understand the key success factors for overcoming a range of marketing challenges and give you the tools to put specific programmes into action quickly and effectively.

1

Position the Company

Successful technology companies have a clear view of their position in the market. AT&T sees its role as 'connecting you to your world, everywhere you live and work'. Cisco supports the 'human network'. IBM, once the global leader in mainframe computing and a pioneer of personal computing, now positions itself as a services company committed to 'building a smarter planet'.

What these broad positioning statements do is give your company the freedom to market a wide range of products, services and solutions within that framework. New products, acquisitions and mergers all make sense if they support the broad strategy. Compare that with a company that positions itself as a supplier of network components, security solutions or mobile devices. Those are still broad categories, but they impose boundaries on that company's market opportunities.

Clear positioning helps you create a sustainable competitive advantage by focusing on strengths that are hard to match and clearly differentiating your company from competitors. Positioning is also important from an internal perspective. It ensures that everyone in your company aligns their activities to the overall strategy. Positioning tells your employees where the company wants to be and how it will get there. That clarity and consistency gives direction and focus to your product development programmes, investments, acquisitions and other critical activities that influence success.

With that focus, you can identify and prioritise the most important market opportunities to pursue and, just as important, recognise activities that don't fit the strategy and discard them, rather than waste scarce marketing resources.

It's also essential to communicate your positioning to external stakeholders such as customers, supply chain partners and investors. Customers and supply chain partners incorporate your strategy into their own future plans, particularly where it relates to product development or manufacturing capacity. Investors use the information to make decisions about their future commitment to your company.

Corporate Vision

An important medium for communicating your positioning is the corporate vision. It influences the way customers and prospects perceive your company as a supplier or potential business partner by helping them understand your direction, mission and values and recognise your strengths and achievements. By consistently communicating values of stability, flexibility and innovation, companies like IBM, Dell and Sony have maintained their strength in the marketplace.

A corporate vision sets out the position you seek to achieve in the future. Examples include these: to become the leading provider of healthcare data management solutions, to achieve market leadership in software for the financial services sector or to achieve leadership through innovation in networking. The vision provides a framework for marketing operations that help you achieve the vision, communicate your vision and encourage customers to experience your company's vision by buying your products or using your services.

Technology event management company RSA Conference organises a series of regular events for information security professionals. The company positions its conferences as events with a collegiate atmosphere that represent an influential, trusted resource. Its marketing communications aim to earn the trust and respect of industry colleagues and customers by demonstrating how its events empower the target audience. Communications demonstrate how the company understands the preoccupations and priorities of its audience, speaking in a conversational tone, addressing audiences directly and speaking on their level as a respected colleague.

The communication strategy positions RSA Conference at executive level, demonstrating how the collegiate atmosphere enables delegates to collaborate informally and network with top industry experts as well as their peers to find new solutions to the most important issues their companies face. Positioning like this is particularly appropriate for conferences, workshops and other customer events. The ideal conference, according to RSA, would have the feel of a reunion, characterised by involvement and a sense of being among old friends. The event would be productive and collaborative, with shared information and involvement by people who are glad to be participating, and who enjoy working together in a spirit of fun and liveliness. The conference should present delegates with the latest products, identify the most important issues, provide ample networking opportunities and allow them to get involved with the industry's leading innovators in a forum that features a high level of interactive participation.

To establish your vision, you need to examine the marketing environment you operate in and identify the type of company you need to become to succeed in that environment. The vision of the company you need to become drives your business, financial, human resource and marketing objectives and establishes the priorities for those objectives.

If your firm's vision is to be a market leader in a specific sector, this will drive a number of marketing objectives. These might include increasing market share from x per cent to a market-leading y per cent over a period of three years, building a distribution network that provides comprehensive market coverage, or creating a product portfolio of best-in-class products that offer benefits competitors cannot match.

The marketing objectives that will deliver the vision drive a series of marketing strategies. To achieve the objective of gaining leading market share, the strategies could include winning x major accounts from competitors or strengthening relationships with y existing customers that could possibly move to competitors. To create the best-in-class product portfolio, strategies might include developing a range of new higher-performance products, adding products from other companies through collaboration or acquisition, or upgrading existing products.

Alignment of Vision and Brand

Marketing communication plays an important role in achieving your vision and aligning it with your brand. Your brand is the sum of everything that your customers experience about your company – your products, packaging, delivery, customer service, marketing communications and the attitude of your employees. It's essential that you align your current branding with your future vision so that you can begin to change customers' perception of your company over time. The elements of your vision should appear in product communications, advertisements and other communications. By communicating your vision to customers, you can demonstrate that you will continue to be a viable long-term business partner. Customers can use your vision to plan their own future developments, for example by incorporating your new products in their own long-term portfolios.

Here's an example of the way a network company used its vision to reposition itself.

Network companies provide internal and external infrastructures for customers to run business applications and transport their voice, video and data traffic between different users and locations. They aim to grow by carrying increasing volumes of traffic cost-effectively and stimulating the development of new applications that will generate revenue from additional network traffic. A key objective for network companies is to reposition themselves as strategic business partners rather than equipment suppliers. They do this through thought-leadership programmes aimed at business and financial decision-makers, as well as their traditional target audience of technical decision-makers. White papers and presentations explain how the right network can help other organisations collaborate more effectively, improve customer service, reduce communication costs and streamline business operations throughout the supply chain.

To deliver the strategic services, network companies must first achieve the objective of building an end-to-end network with national or global coverage. An end-to-end network enables their customers to communicate quickly and easily with any other location connected to the network. When a network company has built a single network that reaches locations worldwide, it can offer a single-source service to customers who need to do business on a global scale.

Network companies can achieve their revenue goals by offering a wider range of services over the same network. Service providers originally built separate networks to carry voice or data traffic because of their differing technical requirements. The latest multi-service networks are capable of carrying voice, data and video traffic on the same network. Network companies with multi-service infrastructures can offer customers a single source for all their traffic, increasing revenue and strengthening customer relationships.

As networks grow more complex, many organisations find it difficult to manage the network using their own resources. Network companies can reinforce their position as strategic business partners and take advantage of a new revenue opportunity by developing a range of support services. Network companies can deliver those services using their own staff or work with a quality-controlled service partner.

When customers run more applications on a network, traffic and revenue increases for the network company. An important objective for network companies, therefore, is developing new applications internally or collaborating with independent developers. AT&T, for example, introduced a collaboration program that enabled developers and technology start-ups to partner with the company and accelerate new application development.

Given the rapid changes in technology, network companies must develop a strategy to ensure that their vision stays ahead of change. Cisco's corporate overview explains how the company aims to capture market transitions, such as the move to a single network for all traffic. This enables network companies to help customers evolve, as their industry and needs evolve.

Strong positioning provides a clear advantage in the marketplace when it creates the impression that your company and its products have few credible substitutes. When your company has established recognition and trust in the marketplace, it is likely to obtain more sales opportunities. You will gain the status of a pre-approved supplier in any purchasing decisions, increasing your chances of success. Corporate positioning also helps to determine the prices you can charge and the resulting profit margins. Analysts believe that people want to buy more from a company with a good image and they will pay more for its products. Developing a positioning strategy forces your company to consider what it needs to do to be successful. Your strategy should define the target market where you compete and the challenges the market faces. It should highlight the solutions your company offers to those challenges and the key benefits that differentiate your company.

To use a corporate positioning strategy effectively, you need to align your products, customer service and your people with the strategy. Simply using the positioning strategy to develop advertising slogans or corporate mission statements is not sufficient. By working with all levels of staff, you can ensure they are aligned with the positioning strategy and understand the part they can play to achieve the position. You should also develop a series of key messages based on the positioning strategy and share them with all customer-facing staff. This will ensure that staff communicate consistently in every customer contact and reinforce the corporate positioning.

Key Messages

You must present your capability in a way that is relevant to your customers. Start by assessing your strengths and weaknesses to identify areas where you may need to improve your own performance. Identify the strengths and benefits you bring to your customers and research what they need to know to evaluate you as a supplier. Your positioning statement should include the following key messages: commitment to your business, your future direction, market experience, technical expertise, management capability, company resources and financial stability.

Customers will feel more confident when you can demonstrate that you have invested in the future growth and development of your business. Describing your long-term plan for the business and including a statement of direction shows how you plan to develop the business over the long term. Timescale is important, so describe your plans for growth and outline your vision for the business in, say, five years' time. By including information on your future product or service range, you can help customers align their own long-term plans. Outline your plans for expansion so that customers understand how they will be able to take advantage of future facilities or capabilities.

Your track record helps to build credibility. Case studies help to explain why you are capable of understanding your customers' requirements and show that you have already developed successful solutions in that market. The cases should focus on specialised market knowledge that is of interest to your customers and provide examples of technical innovation or leadership to prove your capability. You can reinforce the innovation message by detailing your annual expenditure on research and development, your record in new product development and your technical and research resources.

Strong management is important to customers. Show that you have the resources to manage your business effectively by outlining the experience of your management team and list the key facts about your company: your size, number of employees, location, turnover and profitability, national or international network and your infrastructure.

Customers want to know that they can rely on your company to continue providing them with the same high standards of service over the long term. That means making sure that your customers have a full understanding of the financial structure and performance of your company. You should describe any major investment programmes that you are carrying out and provide your customers with regular information on your financial performance. You also need to demonstrate that your production resources are capable of handling increasing volumes of business.

Positioning messages provide your company with a clear framework for communication. They set out the facts that differentiate your company from competitors and communicate benefits that are important to the target audience. Positioning messages are crucial to the success of marketing communications.

Repositioning a Company

Key messages like those described above can also help to reposition your company in the marketplace by changing customer perceptions. Korean electronics manufacturer Samsung, for example, moved up the value chain by changing its image from supplier of budget products under various brand names to manufacturer of innovative, high-quality consumer products, grouped under a single powerful brand. Repositioning your company can be important in a number of business scenarios:

- Your company has undergone significant change.
- Research shows that customers are not aware of your company's key strengths.
- Your company is entering new markets and there is low awareness among potential customers.
- Research shows that the company has a poor reputation in a number of areas important to its success.
- Purchasing decisions are influenced by executives who are not aware of the company.

- Your company is trying to build partnership relationships and customers need to know that you are capable of providing stable supply.

Before you consider repositioning, carry out research to assess how the target audience currently perceives your company. Identify the target perceptions that will persuade the target audience to hold a positive and favourable attitude towards your company and audit existing communications to identify current messaging. Comparing your current position with target perceptions helps you set objectives for developing key messages.

This is the management summary of an audit of customer perceptions:

- Contact with the customer at all levels is less than professional. According to the customer, the company does not understand its business and its products, and does not communicate its future strategies.

- There is a legacy of poor reputation which has largely been overcome by increased product reliability, but the image persists in the minds of the customer's senior management team.

- The company is perceived as offering lower quality and lower performance than competitors.

- The company is seen as losing ground with important decision-makers.

- The company is identified more clearly than competitors with specific product lines, but is not rated most highly as the potential supplier of those products.

- The company's major weakness is perceived as its narrow product line and lack of expertise in certain areas.

An audit like that identifies key areas for improving communications performance and it is essential that these messages should be communicated consistently in every form of contact with your customers. Significant developments such as new investment programmes, acquisitions, changes in management, expansion programmes or new product launches are of major interest to your customers. Take the opportunity to update everyone on progress.

If your company has a poor reputation in the market, positioning messages can help to overcome that. It can help to correct wrong perceptions, but, if the perceptions are based on poor performance, the focus should be on improving the performance itself. Trying to mislead the market can be dangerous.

If, on the other hand, your company's products have an excellent reputation in the market, why should you worry about communicating messages about the company? Success depends on more than a good product range. Your company may have excellent products but a poor delivery record. If demand is growing, customers will ask if you have the capacity to meet new levels of demand. If your company is not making good profits, customers will ask whether it can invest for the future or even survive in the long term. These are good reasons to keep customers informed about your company.

Communications Framework

To ensure consistency, publish a framework for positioning messages. This is the equivalent of a corporate identity manual for messages. Issue the framework to all customer-facing staff and teams from marketing agencies and create message templates for standard text used in press releases, presentations and marketing publications. Set up a procedure for reviewing and approving the key message content of all communications material before publication. Carry out continuous research to measure the effectiveness of the key message programme and compare perceptions with the baseline at the outset of the programme. By using the framework to ensure consistent use of key messages in all communications, you can improve the results from marketing programmes.

Positioning statements stress the positive aspects of your company and seek to correct any misunderstandings. By looking at the factors that are important to your customers, you can help to build their confidence and trust. Here's an example of how a company could shape up with a positive reputation in the eyes of its customers:

- The company is financially sound. It has the stability and financial strength to grow in line with our own increasing demands.

- The company is well managed. It has a dynamic management team committed to customer service and it attracts talented executives in every discipline.

- The company has a clear sense of direction. It is committed to market leadership and technical excellence. Its long-term product plans will help us to improve our own business performance.

- The company is a market leader and sets the standard for the rest of the industry. They are attracting customers who demand the highest standards of service.

- The company has an exciting new product range. It is clearly capable of innovation, and partnership with the company will clearly benefit our organisation.

- The company cares for its employees. It spends a great deal on training and personal development and that is reflected in a commitment to quality.

- The company has an excellent investment record. Its production plants are highly productive and that means quality production with good costs.

Most customers would feel confident in a supplier like that. They could trust their supplier not to let them down on quality and reliability and they could expect the supplier to help them push their own business forward. That's the kind of reputation you can aspire to.

The recent economic downturn has put an even greater focus on stability. When customers are making major purchasing decisions, one of the first questions they ask is 'Will that supplier still be in business in five years' time?' That's important because the customer's own business depends on working with suppliers who understand their needs and can meet their requirements over the long term. So confidence really matters.

Reputation Management

Although you can take positive steps to build your credibility, the growth of social media makes it harder than ever to control your reputation online. Blogs, product review sites and social networking tools give anyone with Internet access a platform for voicing their opinion on your products or your business. The increasing number of online channels also makes it difficult to monitor what people are saying. If they are making negative comments this can damage your reputation, so it is essential to track comments, respond positively and take any necessary remedial action.

As part of your positioning strategy, you must manage your presence in social media. Identify the channels where people are talking about your business and monitor search engines to find references to your company, product or brand names. The references could be in press, radio or television items, blogs, social media sites, forum posts and websites of other companies. Check the search engine rankings to see how high any negative references appear. You can counter unfair or damaging online references content by publishing positive content that will help to push any negative material further down the rankings and reduce its search engine visibility and impact.

It's important to respond to any criticisms or suggestions. Acknowledge the author of the comments and thank them for highlighting any issues. Respect any valid criticism and promise to deal with the issue. Describe any remedial action you plan to take with a timescale for improvement. Identify any individuals that you feel warrant a personal response; this can help to create goodwill and that may result in positive feedback and comments in the future.

Building dialogue on the relevant channels is an important part of managing your reputation. Post your own comments on blogs and forums to communicate your opinions on hot topics and demonstrate your willingness to engage. Set up forums, communities or review facilities on your own website to encourage further dialogue and take any necessary remedial action. If critics highlight product issues, identify product development tasks to improve performance or usability. If service issues are the problem, review your customer service operations and identify opportunities for improvement. Prioritise any problems that are generating high levels of comment.

If you face serious problems from online critics, appoint a firm specialising in reputation management. The public relations role has increasingly become about managing and interpreting relationships, encouraging listening and engaging in dialogue in order to protect and maintain reputations.

Summary

Positioning your company clearly and managing your reputation are important to marketing success. A strong reputation ensures that you will always be considered when companies are looking for suppliers. The ultimate indicator of successful positioning would be category ownership – a situation in which your company is synonymous with a technology solution.

2

Understand Your Customers

Companies buy technology to solve their business problems and overcome pain points. The success of your marketing programmes depends on a clear understanding of those issues. The emphasis on business drivers is borne out by industry surveys of IT purchasing plans which indicate that companies emphasise business growth, but continue to be vigilant about IT costs and operational efficiencies. Three of the top four business priorities in recent surveys were growth-related: increasing enterprise growth, attracting and retaining new customers and creating new products and services. The top technology priorities supported both growth and cost efficiency: cloud computing, virtualisation and mobile technologies.

Business Drives Technology Purchasing

Business objectives are an important factor driving IT purchasing decisions. IT executives and managers work closely with business decision-makers to identify where technology can provide a competitive advantage. They also consider Return on Investment when they are making a business case for technology initiatives. Purchasing technology for its own sake is no longer an option.

As an example, growing businesses need additional IT resources to support product development, communications and customer service. When the company needs those additional resources quickly to take advantage of business opportunities or reduce time to market, this puts the IT management team under additional pressure. In response, many companies are considering cloud and virtualisation as a means of expanding resources quickly. As well as responding to business needs, the IT management team faces the challenge of finding people with the right skills to implement and operate technologies such as cloud, virtualisation and networking. Security, in particular, remains a major challenge for IT management. The two most important tasks are to protect the IT infrastructure so that it remains available for users and to secure the company's data. However, industry figures indicate that many companies had experienced cuts to their security budgets. That's despite the fact that a high number had experienced at least one information security breach in the previous 12 months.

Changes in the IT landscape are also complicating IT security management. As companies extend their network access to supply chain members outside the corporate security firewall, the IT management team must try to maintain security over external devices and networks that they cannot control. Another internal challenge is the growing use of consumer IT devices such as smartphones, netbooks and social networks in business. An increasing number of businesses are allowing users to suggest which smartphones to buy. These devices and networks are not subject to corporate security policies and are therefore increasing the overall security risk.

Pain Points

By understanding customers' pain points like these and developing solutions to overcome those problems, you can differentiate your company and build close working relationships with key decision-makers at all levels. Technology purchasing decisions are typically made by a team of business, technical and financial executives and managers. It is essential that you build relationships with each group and mirror their structure in your own organisation. Although it is your sales team who will make the initial contact, they may not have the experience or the authority to discuss strategic business issues with a senior management team. By involving your own senior executives in the negotiations, you show that customer's business is important to your company.

Elevating products to a strategic business level can also help you differentiate your company from competitors, charge higher prices and build effective relationships with customers. Strategically important products have a number of characteristics:

- Essential to business success;
- Added value;
- Support the core business;

- Speed up business processes;
- Enable new product or market developments;
- Improve competitive performance;
- Reduce the cost of key business processes.

Product development should therefore be focused on customers' business needs. Although most companies carry out research before development, the research may not provide the detailed input that is essential. Product development may also be driven by technology, with no clear market focus. The more your customer depends on your product, the more likely it is to succeed, so involving customers can pay real dividends.

The new products you develop could enable your customers to improve their competitive performance, so it is important to understand their markets. Tell customers about your product plans and ask them for input to your development process. By building a detailed picture of their markets, you can align your own plans with them, and develop products that are tailored to their needs.

- What are their main markets?
- What is their position in the marketplace?
- Who are their main competitors?
- How are their products regarded in the marketplace?
- What are the key success factors in the market?
- What are the long-term product trends?
- What new technical developments will be needed to succeed?
- Could innovation by you help your customers to succeed?
- Are your customers considering entry into new markets?
- Do you have product development plans that are relevant to the new market?

It is equally important to understand your customers' business strategies: their corporate direction and key objectives, and how they aim to succeed. By aligning your product development objectives with theirs, and showing how your products or services can help them to achieve their strategic business objectives, you can ensure that your new products will be successful.

Collaboration Provides Insight

Your customers can also make an important contribution to your new product programme, either through recommendations on direction, features and performance, or through evaluation. Either way, their contribution can ensure that your new products meet real market needs. Involving customers in new product development can result in more targeted products with a greater chance of success. Customers can also make an indirect contribution when they use an online build-to-order system such as the Dell personal computer model where customers effectively 'build' their own model. Allowing customers to configure existing products gives valuable insight into real needs and preferences – information that can feed into future development programmes.

If you are planning a new product or redeveloping an existing one, ask your customers for their views on the existing product and what they would like to see in a new one. By explaining your plans and involving customers in product development, you can strengthen relationships and provide a service that is mutually beneficial. Questions could include:

- How can we improve the current product?
- What problems need to be overcome?
- What new features would customers welcome?
- Do the plans represent an improvement?
- Would customers make greater use of a product that included the features they have highlighted?

You can encourage feedback and build a sense of community by setting up a user group. The user group would operate as a forum for discussing issues of mutual concern to customers such as quality, performance, standards, future developments and customer concerns. The group would include representatives from your own company and from a cross-section of your customers. Comments from the user group provide valuable feedback on current performance and help to identify needs that can be met through new product development. However, there are many examples of companies who have set up user groups in response to a crisis and then failed to use them. This can be frustrating for customers and wasteful for the companies. User groups provide a valuable perspective on products and service, and their feedback can provide real benefits for the product development process.

As well as asking for input, you can also involve your customers in trials of new products. Customer evaluation, or beta testing, is well established in the software industry. Customers test new products or upgraded versions before they are released to the market. They identify any problems in using the software, providing valuable feedback on product performance. Any problems or practical improvements can be incorporated before final release of the new product. Although there are important benefits to this type of collaboration, there are also risks. Firstly, the customer may be extremely disappointed with the product if quality is poor. Secondly, there is a risk that competitors could find out about your plans indirectly. The quality issue is one that you should deal with: if a product is not right, it should not be given to customers in any form – it is not enough simply to promise future improvements. The security risk of a leak to competitors can be minimised through disclosure and confidentiality agreements, although these provide no real guarantee. However, the advantages of involving customers outweigh the risks, so evaluation is worthwhile.

Another important practice is to pre-announce new products. For example, a software company will set a number of release dates during the coming year when it will release new versions of products. The company outlines the new products and gives customers the opportunity to provide input to the development process. The major benefit for customers is that they can align their own business development plans to the release dates. They might, for example, postpone a particular project until the latest version of a product is released in six months' time. However, some companies have put themselves under unnecessary pressure by trying to meet a series of pre-announced release dates. The schedule may not allow proper time for development, resulting in failure to meet the date, or the release of a product that is not ready. Both are potentially damaging.

Product development can be a joint initiative where you work closely with specific customers to develop products that meet their specific needs. This approach is a valuable one where:

- Your customers have developed partnership sourcing to take advantage of your technology.
- Your customers have technology and technical skills that complement your own, and a joint project can produce more effective results.
- You want to strengthen relationships with key customers by working in partnership on joint development projects.

There are different levels of collaboration. Some may involve regular meetings to provide input and review progress. These meetings can be held on site or remotely, using videoconference links. In some cases, customer staff may work alongside the supplier team for all or part of the project. Secondment like this can provide other benefits for the customer by improving the technical knowledge of their staff.

Customers can make a direct contribution to the product development process by getting involved in the ways we described, but there is also an indirect form of involvement where customers choose their own product specification. When Dell first allowed customers to configure personal computers online, they revolutionised e-commerce. Buyers were no longer limited to standard, packaged solutions. Now, customers expect manufacturers to offer products configurable to their needs and to deliver them quickly.

Allowing customers to configure their own products gives them greater choice and enables you to gain a better understanding of customer preferences. The Dell model gives buyers a choice of standard features with a simplified selection process. Each selection is priced as an addition or subtraction from the base price option and may include free upgrades on certain features. The model also includes information on the benefits of choosing different features with comparisons between alternative choices. Customers can visit other parts of the website to read product reviews and customer feedback. They have the option to download brochures and can edit their choices at any stage before purchase. If they are not buying immediately, customers can create different versions and save them for comparison or later purchase.

This is a proven commercially successful model that can be applied to technology products and services that are available with options. However, some commentators question the viability of the configurable business model given the cost issues and logistical problems arising from customised production and individual product fulfilment. They believe that it may not be suitable for products that return small profit margins per unit, particularly mass market products where price is the major deciding factor. Customisation is not suitable for products that have few options or products that are low value.

A Complete View of the Customer

Understanding customer needs and developing customised solutions is critical to building the right relationships. However, the customer's experience of your business goes beyond products and services, so it is important to get a view of the bigger picture to ensure that you deliver the right level of service at every opportunity. Building a 360-degree view of the customer takes understanding to a new level.

Most companies have detailed customer records, but how many are aware of all the customer interactions that take place? Very few have a 360-degree view of the customer – that is a single view of all customer interactions. Getting a 360-degree view is essential if you want to optimise marketing, communications, customer service and product development. Used effectively, the 360-degree view strengthens customer retention and can increase customers' lifetime value and profitability. Looking ahead, if you have a 360-degree view, you can tailor products and services to meet your customers' long-term needs, confident that you have a complete picture.

The 360-degree approach underpins a number of important customer-service initiatives:

- Development of measurable goals, objectives and tactics for every point of interaction customers have with your company;
- Establishment of company-wide business metrics, customer measurement, tracking and reporting processes for all sectors, product lines and points of contact;
- Customer interaction guidelines that span every part of your company that has customer contact.

Customer information is often scattered across several business systems making it difficult for customer-service representatives to gain a 360-degree view of the customer when they are making contact. As a result, staff waste valuable time navigating several applications to piece together a single view of the customer. They may also miss opportunities to make inbound customer enquiries more productive because they lack customer intelligence, such as a customer's disposition to buy certain products. A customer relationship management (CRM) system can overcome this problem.

The information in a CRM system is designed to ensure employees can offer a prompt response and a personal, efficient service to customers each time they contact a company. An effective CRM system centralises all relevant customer information and makes it available to any employee dealing with a customer.

A basic customer profile should include name, contact details, product and warranty information, purchase history and service records. CRM systems take these profiles a stage further and include up-to-date information on all customer interactions with the company. A CRM system can capture contact information from any communication channel, including telephone, email, website and fax. That information provides insight into how customers are contacting a company, browsing a website, acquiring information, ordering products, placing service enquiries or making complaints. It brings together all that information to build a detailed profile and shows how well service, marketing and communication channels match customers' activities.

CRM systems incorporate business rules that help agents provide the right level of service to individual customers. The rules are based on factors such as the importance of the customer, purchase preferences, credit limits or contract service levels. The business rules guide the actions an agent can take for different types of inquiry and also include prompts to help the agent offer additional products or services relevant to the customer. As well as customer information, customer-service management systems should also include details of employees with the skills to deal with specific requests or problems. This is particularly important in support and help desk operations where customers need expert advice to resolve problems quickly. The information in customer-service records provides a basis for business analytics applications that enable companies to analyse customers' past behaviour and anticipate future trends.

Networked solutions provide an efficient and effective method of creating a 'single view' of the customer or a single view of a process. The system integrates your other business systems at the desktop of customer-facing staff, providing them with a comprehensive view of the customer that can be used to simplify access to customer information and dramatically improve productivity, morale and, most importantly, your customers' experience.

A 360-degree solution enables you to consolidate customer data such as transaction history, payment history or other information from separate applications and deliver a simple view that any member of staff can use to provide the information a customer is requesting without having to navigate a maze of screens. Your customers want information and services fast. Providing your staff with a single view of the customer can have a significant impact on their ability to deliver superior customer service. By reducing the amount of time and frustration it takes to switch between multiple applications to handle calls, you're giving staff greater flexibility, enabling them to spend less time interacting with systems and more time interacting with customers. Having full knowledge at their fingertips means that staff can close each enquiry successfully without having to refer to other people or other systems.

The ability to deliver a high-quality customer experience using a 360-degree view of the customer can be of major strategic value to your company. Having a single view of the customer right at your staff's fingertips means that you have a greater opportunity to cultivate a more meaningful relationship with that customer, either solving a problem or offering additional products or services that generate additional revenue. The 360-degree view enables your staff to increase customer satisfaction by engaging in knowledgeable conversations with the customer, confident that they have full up-to-date information available when they need it.

The 360-degree view doesn't just offer a consolidated view of the customer, it also provides staff with access to information needed to meet customer-service demands and support Web-based customer self-service. The system can also guide staff through processes such as cross-selling interactions, which allows you to introduce best practices into every customer interaction. The result is that customers will enjoy a consistent experience throughout your company.

This level of consistency also provides important productivity benefits for your sales and customer-service operations:

- Improved customer interaction and call quality;
- Reduced call time;

- Significant reduction in data entry errors;
- Reduced training time.

Differentiate Through Service

Customer service and sales operations based on a 360-degree view give you the opportunity to provide your customers with a one-stop shop – a single point of contact for all your products and services. This can be particularly useful if your company has a number of product lines sold through different departments or divisions. Customers don't want to spend their time being transferred between departments or trying to work out for themselves whom they should contact. Ideally, any member of your customer-facing team should be able to deal with any type of customer enquiry. It may be that certain enquiries have to be escalated or transferred to specialists, but this can be handled seamlessly without inconveniencing the customer.

Increasingly, customers are making buying decisions based on more than the four traditional factors of product, place, price and promotion. They are also influenced by variables such as service quality, recognition and efficiency of customer support. With a 360-degree system in place, you may be able to strongly differentiate your company in these areas and achieve high levels of customer satisfaction.

The 360-degree view is an integral element of CRM. Relationships with customers are not possible without understanding who your customers really are, what they value, what they want to buy and how they prefer to interact with and be served by you. Moving your customer relationships forward requires a deeper understanding of customer characteristics and behaviours, such as your customers' purchases and related spending patterns, their current potential value to your company and other characteristics.

As well as providing a single view of the customer, the consolidated information available from a 360-degree approach can also help you segment your customers into segments. By sorting customers into groups that differ in purchasing behaviour, annual spend, product preferences, demographic and other characteristics, you can customise product offerings, service and support systems, order processing, marketing communications and other significant customer interactions.

This is important because not all customers are equal. In most businesses, a small number of top customers contribute the largest proportion of revenue and profit. Therefore, you may wish to ensure a positive experience for your high-value customers across all parts of your company. The system should allow you to flag these customer accounts for special care at the point of contact, directing their inbound calls and emails to senior account managers, providing priority service and making special offers. At the same time, you may wish to direct less profitable customers to use lower-cost automated services or self-service web pages.

Creating a single point of contact for all customer enquiries is an important strategic objective, but to achieve it, you may have to overcome internal resistance. The 360-degree approach replaces separate functions such as call centres, help desks, customer-service departments and sales offices, as well as individual account managers. People in these areas may feel that they have a strong relationship with certain customers and may be unwilling to give up 'ownership'. You need to demonstrate the benefits of the 360-degree approach to all customer-facing staff and win their commitment to the new approach.

The 360-degree approach is based on networking technologies that integrate information from separate departments and IT systems. However, it is the people who interact with customers who deliver the real benefits. They must be committed to the single point of contact approach and they must be able to use the full facilities of the system so that they are able to deliver prompt, quality service and recognise opportunities to win new business.

Summary

Focusing on customer pain points and building a complete picture of their interaction with your company builds a level of understanding that translates into products with strong market appeal and the basis of successful customer relationships.

3

Develop a Long-Term Product Strategy

Technology markets can prove difficult if you do not have a clear point of differentiation. Open standards and common components have turned many products into commodities, while shrinking product development cycles make it difficult to sustain a competitive advantage. In the personal computer market, for example, virtually all vendors use the same operating system, semiconductor and display component suppliers. Comparative performance at any price point is similar and differentiation is difficult.

Rather than resort to price competition, the challenge is to achieve long-term product leadership by delivering value to the customer. Companies that are focused on product leadership align their complete organisation structure behind delivering value. The structure features a strong leadership team heading cross-functional teams that are focused on results. Product leaders also extend their organisational structure to include strategic partners.

Product Leadership

Achieving product leadership is a strategic decision that may involve significant changes to your company structure. When telecommunications company Nokia reported its plans to challenge for leadership, the key organisational elements of the strategy included:

- A new leadership team and organisational structure focused on speed, results and accountability;
- A marketing operation concentrating on capturing volume and value in growth markets;

- Strategic partnership to acquire a product platform for its smartphone range.

Product leadership does not depend on a single individual or department; it takes commitment from the whole company. Success comes from aligning the resources of the whole company to meet the changing needs of your customers. That means your senior management team must set out a vision of product leadership, explain the role of different departments in achieving leadership and commit resources to any necessary training.

Cross-functional teamwork is essential to achieving product leadership. Marketing specialists provide the customer insight and research to define customer requirements and translate those into a product specification. Marketers can also advise on the development priorities so that new products meet customer needs, rather than internal requirements. The product team should also include members from engineering, research and development and product management.

Partnership with suppliers and other external organisations may also be important to achieving product leadership. Partnership can add value to your product leadership programme by providing access to skills, technologies and other resources that you do not have in your own company.

Product Management

A key resource in the drive to achieve product leadership is the product management team. Product management is the process of managing your products throughout their life cycle. Product managers have to meet a number of general objectives, including developing a product plan in line with your corporate strategy, maximising revenue, profit and customer satisfaction, delivering new products on time and on budget, and achieving product quality levels.

Product management has a strategic role and a dual responsibility – to manage the development of new and existing products and to take those products to market within the guidelines of various internal product strategies. From a product perspective, they must consider a company's platform strategy, its future product roadmap strategy and the product life cycle. Product strategy is driven by marketing strategy, aligning customer needs with product features.

A key objective for the product management team is planning and developing the specifications for a range of products or a product portfolio that meets your long-term strategic plan. The strategic plan may require development of new products to meet the needs of new market sectors or improvements and extensions to the current product range to increase share in your existing sectors.

Product managers must ensure that products meet or exceed customer needs. That means working with your customers and utilising their feedback to ensure that your products are easy to use, simple to maintain and capable of delivering value to your customers. Enhancing the user experience helps product managers to meet the important objective of maximising customer satisfaction. Product managers have to meet time and budget objectives. To meet the demands of the market and counter competitive pressures, they must be able to bring product development programmes to a conclusion on time and on budget.

The roles of product management and marketing management share similar objectives, which are to maximise revenue and profit by meeting customer needs. Some companies combine the role of product manager and marketing manager. In others, product managers work closely with marketing managers, using research from the marketplace to plan and prioritise product development programmes and briefing marketing teams on the benefits of new products so that they can develop effective customer communications.

The ultimate measure of product management is commercial success. Although product managers may not have specific profit and loss responsibility, they must recognise the importance of financial results. To achieve commercial success, they need to coordinate the activities of other specialists in your company who contribute to the product's performance in the marketplace, including representatives from marketing, research and development, finance, engineering, manufacturing, business strategy and product management. Bringing together people from different disciplines helps to raise awareness of the importance of new product development throughout the company.

An integrated team like this helps your company to identify and take advantage of new market opportunities. The team achieves this by carrying out market and competitor research, analysing market requirements, identifying new product and service needs, and supporting sales, marketing and product development activities. Business development teams should also identify opportunities to improve service to your existing clients as the basis for increasing revenue.

Select team members with the knowledge and experience to contribute to business development. Your team members should have five important characteristics:

- A successful track record in business development;
- Skills to gather and analyse information;
- Ability to recognise opportunities and qualify prospects;
- Communication skills to deal with contacts at any level;
- Leadership skills to change the way a company conducts business with customers.

It's important to build a balanced team with the skills to support all aspects of business development. Appointing a senior executive ensures that business development activities align with overall corporate strategy, while representatives from marketing bring customer focus and market understanding to the team. You should also involve members of the sales team who deal directly with clients and prospects. Ask a member of finance to manage funding for business development activities.

You can improve opportunities for success by introducing best practice and consistent processes for business development. The Product Development and Management Association offers certification in new product development principles. The programme helps to develop a structured approach to new business and new product development using standard terminology and processes. This approach simplifies communication and ensures that all team members and contributors understand their role in business development. To ensure progress and deliver results, set measurable targets for the team, establishing metrics for targets such as revenue growth, increase in market share or new account openings. Motivate the team with incentives related to the target and communicate new business achievements throughout the company to recognise the contribution of the team.

A new product team focuses your company on innovation as a way of improving your competitive advantage and building future revenue sources.

It's important to recognise that innovation can come from many sources, not just product development specialists. Setting up a cross-functional product team helps you to source ideas for improving existing products, as well as identifying opportunities for new products. A broadly based new product team helps to ensure employees have a common level of understanding and knowledge when they are involved in new product development.

Your product development programme must focus on new products that meet customer needs – innovation for its own sake does not guarantee success. Use market research to identify the strongest trends in customer needs and check the websites of patent licensing agents. They list companies looking for inventions in different sectors and that information can highlight gaps in the market.

You can encourage innovation inside your company by setting up facilities on an intranet for employees to post suggestions. Set up a formal product development programme to identify ideas that represent strong market opportunities. *Microsoft Secrets,* by Michael Cusumano and Richard Selby, describes the culture and processes that enabled Microsoft to bring so many innovations to market. According to the author, the key elements in the company's success include:

- Finding smart people;
- Organising small teams of overlapping specialists;
- Pioneering mass markets;
- Focusing creativity;
- Improving through continuous feedback.

Innovation can come from outside your company as well. Contact suppliers, business partners, research organisations and universities to source new ideas or to assess the commercial potential of projects that they are working on. You can also license existing products or technologies from other organisations to expand your product range. Rebrand the products or offer them to consumers as own-label products.

Pharmaceutical research firm Biofocus, for example, found that location on the Cambridge/Essex border was important to the success of its collaborative product development programme. The company is close to Cambridge University and a cluster of potential partners, including pharmaceutical manufacturers, biotechnology companies, patient organisations and charities that focus on developing solutions for specific diseases.

Evaluate any innovations or suggestions to see if they provide a unique solution to a real problem. Compare the proposal to existing products and assess whether the differential would create a product that could generate revenue and profit. You can also work with independent product development specialists who have promising ideas but don't have the resources or the funds to turn their ideas into commercial products. To bring a product to market may require design engineering, prototyping, production, marketing and customer service resources which your company can provide.

Product Development Process

New product teams should meet regularly to ensure a flow of ideas for product development and to ensure that other employees or third parties have a clearly identified channel for submitting new product proposals. The new product team has a changing role at different stages of the product development process:

- Generating ideas;
- Evaluating proposals;
- Testing concepts;
- Preparing a business cases;
- Product development;
- Test marketing;
- Product launch.

A new product team with members from different functions provides a broad perspective during status and review meetings. This ensures that the project meets business, technical and financial objectives at each stage.

The initial phase focuses on generating ideas for products that meet customer needs and offer solutions that provide greater benefit than existing products. It is important to carry out research into customer needs and competitive products to generate relevant ideas. The initial ideas are then subject to screening. The aim at this stage is to identify and filter the ideas that are most likely to succeed commercially.

The team takes the most promising product concepts and assesses them from a marketing perspective. Winning companies fine-tune their product development process to maximise potential market success for new products and use the knowledge and experience to improve subsequent development efforts. The team puts together an outline marketing plan which is then subject to business analysis. A senior management team considers estimated demand and compares potential revenue with projected development, production and marketing costs to reach a provisional profit forecast.

Product concepts that demonstrate a potential Return on Investment progress through detailed technical and production development to create a prototype for test marketing. The test market helps to validate the initial marketing proposals and identifies consumer response to the prototype. The product may then go through further development and testing to arrive at a final version for launch. Feedback from the test market helps the team to prepare a full marketing plan with a particular focus on the launch. In some variations of the product development programme there is a further stage of post-launch review. This can be important in determining the direction of future product development and marketing initiatives.

New product teams should not move seamlessly from one stage to the next. The levels of investment required and the strategic importance of new products make it vital to review projects at each stage. Many companies use a milestone gate review at the end of each phase to assess the project against its business, technical and financial objectives. The review determines whether the project should continue and identifies any risks that need to be overcome. It's not essential for new products to go through every one of the review phases. An agile product development process enables you to introduce a succession of incremental product improvements that create the equivalent of a new product each time. This speeds up time to market and reduces risk in the new product development programme.

Product Platform

Another important technique for reducing development timescales is the use of product platform and product roadmap strategies. Companies who produce different versions of a core product for individual markets and customer segments manage their product designs using a platform strategy. The platform strategy reduces research and development effort and enables streams of products to be launched on the underlying platform. This approach also helps to control development and manufacturing costs when compared with producing a new version for each market.

A product platform strategy enables you to establish standards and processes that other parties can work to when they collaborate with you. In the computer industry, for example, software companies provide independent developers with code and other forms of support so that they can develop additional new products that are compatible with the underlying platform. Platform products help manufacturers to control costs and take advantage of new technologies without major investment.

A product roadmap strategy sets out the company's future direction for product development. The roadmap may be based on a platform strategy, developing future products on a common platform. Alternatively, the roadmap may set out proposals for new products that take advantage of emerging technologies or new, advanced materials. The roadmap gives product managers and other parties involved in product development a set of guidelines that coordinates their individual activities.

A company's marketing strategy sets the requirements for product strategy by translating your customers' needs and challenges into product specifications. In some cases, marketing strategy might identify a need to improve existing products to meet competitive challenges or declining customer satisfaction levels. Marketing strategy might also require new products to meet changing customer needs or new market opportunities. Product managers must take account of both market and product strategies in developing their product plans.

Any product development decisions must also take account of the product life cycle – the stages a product goes through from initial launch through rapid sales growth to a point where sales decline. The four stages are introduction, growth, maturity and decline. Companies invest different levels of marketing or product development funding at each stage to maintain profitability. Product managers may decide to develop new features during a product's growth and maturity stages. However, investing in a product with declining sales is a waste of resources. The emphasis must now be on developing a new product.

Product Portfolio

As part of your product development process, maintain a balanced product portfolio of new and existing products to reduce any risk to the business. A product portfolio is a group of products that have certain characteristics in common. They may be products that are aimed at a specific market sector, for example. Product portfolios can include products with a common price or performance characteristics. It is important to group products in line with the needs of your customer, rather than the internal needs of your organisation.

An effective product portfolio provides a balance of new and mature products, high-margin or high-volume products. It should also include older products that create revenue without requiring any marketing resources. Your portfolio should contain a mix of products that meet and exceed the needs of your customers. Group products in a logical way so that your sales force and other customer-facing staff understand the opportunities for cross-selling products in the portfolio. Products in a portfolio should also create a complete solution for the client. For example, a portfolio might include hardware, a range of application software and services such as consultancy, design, installation and maintenance. Your product portfolio should also align product development with your market sector strategy. Identify market sectors where there are opportunities to sell a range of products. Bring together products and services that you are already selling to the market and products that your competitors supply. Analyse your portfolio to identify gaps in coverage and plan a product development programme to fill gaps in the portfolio. The portfolio provides a useful structure for developing future range extensions and new products that serve the same market sectors. A portfolio strategy enables you to develop products in a controlled way.

It's important to manage the products in your portfolio in line with the stages of the product life cycle – launch, growth, maturity and decline. By assessing the products against life cycle stages, you can identify the areas where you need to focus marketing. That will ensure that products in the launch and growth cycles receive marketing support to improve revenue and profitability.

You should reduce the support for products in decline and operate a product development programme to ensure that you maintain a balanced portfolio.

Financial performance is another important aspect of product portfolio management. Analyse your current product mix to identify the products that provide good profit margins or achieve growing sales in markets important to your business. Carry out customer research to identify areas for improvement or ask customers for a wish list of product improvements. If you monitor feedback on product review sites or customer forums, you can identify current weaknesses or customer concerns. You should also compare your product specifications with competitive offerings and consider eliminating products that have no real differentiation from the competition.

It's important to keep your product mix under constant review and development so that you can continue to meet customer and market needs. Prioritise opportunities to increase sales, create differentiation, improve profit margins or fill gaps in the current mix. Decide whether to retain or dispose of products that offer no scope for improvement. Upgrade the specification of products that currently offer little differentiation from competitors and concentrate on improving features and performance criteria that are important to customers.

If there are gaps in the product mix, fill them by introducing new products. You can develop the new products internally if you have the resources, or work with partners that can supply existing products from their own range or develop the products you need. Ensure that your sales force has the skills and knowledge to sell the new products.

Creating different versions of existing products can increase market coverage. Add or remove product features to offer budget, standard and premium versions of the same product. Computer software is an example of this strategy where the same basic product is available in versions for home users, business users and advanced users, for example. This strategy enables you to reach different market sectors without introducing new products.

Product Line Extension

Another technique for expanding your product portfolio is product line extension – the introduction of a new product that is slightly different to the versions in a company's existing range. Introducing a variation to an existing product line reduces the risk in new product development. Your customers are familiar with your existing product line; provided your line extension offers the same level of quality as existing products in the line and meets specific needs not previously covered by the line, your customers will feel confident choosing the new product.

A product line extension makes an important contribution to customer loyalty. When customers want to buy a higher specification product, they may find that the original supplier does not have a suitable version; instead, they choose a competitor's offering. By extending a product line to include higher value models, you can offer customers the choice of buying from a familiar brand.

Product line extensions can expand a market by giving customers greater choice and providing products that meet customer needs more precisely. You can offer higher- or lower-price versions with different features so that your range meets the needs of customers that you could not satisfy before. Creating different versions of a product is a low-risk strategy for product line extension. A software product, for example, can extend in both directions, offering versions for developers, professional users, consumers and students, with a budget version for cost-conscious consumers. Each version can have additional or reduced features compared to a core version. In the software example, the extended product line may attract users at either end of the scale who had not previously considered the brand.

To plan a product line extension, carry out market research to identify gaps in the market or attractive segments that competitors currently supply. Using the software example, ask customers which features they find most important and plan versions that cover the different requirements.

Extending a product line with new products that use the same brand name can increase the chance of acceptance by existing customers. However, if you are planning to introduce lower-priced line extensions, the question of branding is more difficult. Low price products could harm your main brand if they offer lower quality. In that scenario, it would be better to use a different brand name. If the lower-price product offers the same quality as the main brand, you can position it as an affordable version of your main brand.

Older Products

Some of the more difficult marketing decisions are related to older products in your mix. If products are still generating sales, consider retaining them, but reduce or stop any sales and marketing activity to minimise costs. If sales are declining or very low, drop the product from the mix. Alternatively, offer it to another company that operates in niche market sectors where the product has sales potential.

Failing, outdated products consume management time, sales and marketing costs and production resource, while making a reduced contribution to revenue and profit. To deal with the problem, you need to analyse the reasons for poor product performance, create and implement a product development plan and decide what to do with the failing products. It may be possible to market the failing products in a different way to earn continuing revenue – reducing their price or selling them into new markets, for example. Look at sales over a period of time; declining sales are a clear warning sign that demand is falling or a product is failing. Compare the sales performance of competitors' products in the same market. If the competitive product continues to sell well, the problem relates to the product, not the market. It's important to review the latest products to enter the market; if they are using newer technology, existing products may face further decline.

Analyse the reasons for product failure. Customers' needs or tastes may be changing. New technologies may offer superior performance and existing products may have design or quality problems. Ask customers for their views on the products or check reviews and forums to identify customers' concerns.

Sony Corporation was a pioneer of the video recorder with its Betamax system. However, other consumer electronics manufacturers decided to adopt the alternative VHS system which became the standard for home video recorders, leaving Sony out of the market. When DVD emerged as a consumer technology, Sony participated in the standards group that worked to create a common format for DVD, ensuring that their product would be part of the mainstream.

If you feel that older products can continue to make a contribution, put together a development plan, incorporating the findings from the product analysis. Identify opportunities and priorities for improving or upgrading existing products and consider how new materials, new technologies or design changes could improve the product. Identify and fix any quality problems that affect product performance. Look for new opportunities for outdated products. You could consider offering them as budget versions of the new or improved product. Software companies, for example, offer home and professional versions of the same application program. In many cases, the home version is simply a two- or three-year old version of the latest program. Identify market sectors where price, not product performance, is the important factor.

To avoid damage to profitability, reduce or cut marketing costs and find lower-cost production facilities for the older products. Drop products that have no scope for improvement or alternative marketing. Inform customers of the change and offer them alternative products. You can also offer the product rights to potential buyers or dispose of any remaining stock with a cut price sale to customers or distributors.

Customer Configuration

An alternative to developing a wide product range is to offer customers configurable products on a build-to-order basis. Customers input their own preferences and a factory will custom-make them. Software systems provide you with the tools to assign any number of features to a configurable product, and organise them in meaningful categories that are shown to the user as menus or check boxes. As the customer selects different features, the price builds and changes so that they see the results of their choices immediately

To help customers who are not sure of their requirements, you can suggest a default configuration or standard product that customers can choose 'as is' or edit as they wish. The default configuration and price should reflect good value for money and include the features that have proved most popular with other customers. The product's final price is instantly updated based on your default selections and the choices made by the user.

Customers may need help in assessing the different features and options available. You can build in menus that provide detailed information on each feature as well as comparisons between different features. This can help to simplify the buying process, improve customer confidence and ensure a sale. You can go a step further back and show a list of alternative products based on price range, performance application or other criteria. It is also possible to create feature comparison charts that summarise the functions that different models are best suited for.

Build-to-order systems offer a number of display options that control the way configurable options are presented to the customer. For example, you can choose between radio buttons or drop-down menus, use alternating background colours for tables of specifications, and incorporate pop-up windows to provide the customer with more details about the options available for selection. For more complex sites, you can incorporate interactive 3-D product visualisation that allows customers to rotate, tilt or magnify product images as they change configuration. This feature is useful when design or appearance is important and customers can select different finishes, colours or sizes. Once they have made their selections, customers should have the ability to save the product or service that they have configured to their account so that they can print it, edit it or submit it as an order at any time.

If you plan to sell configurable products online the rule is the easier, the better. The first priority is to make information complete and easy to navigate. You can optimise the process by encouraging feedback from customers through emails, phone calls, site and phone metrics, focus groups, usability studies and online surveys. Responsiveness is key to the success of product configuration. The system should maximise the opportunity for interactivity while minimising the possibility of error.

There should also be a purchase assistance menu that includes payment options, information on tax and shipping, returns, privacy and online security, as well as account, product, shopping, order and technical support information, including the top frequently asked questions. Contact information for sales enquiries or telephone ordering should be prominent.

Once the customer has configured their product, the system should generate order-ready quotes and likely delivery dates. This means that it is important to integrate your website functionality with back-office systems, supply chain and finance partners so that the production and delivery process is smooth.

Global Products

If your company operates in export markets, use product design to match products with local requirements. Companies who follow the 'one size fits all' rule will not succeed globally because tastes, standards, prices, legislation and cultural differences influence customer choice from country to country. Effective product design must reflect global market requirements. Global companies like Nokia know that success is based on a single brand, but with many product design variations.

Companies who operate in global markets can manage their product designs using the platform strategy. That allows them to create a 'core product' with different versions for individual territories. Research into individual country requirements translates the product platform into a market platform plan. Market research identifies customer segments and priorities and maps local product offerings against those requirements. The product design team can then prioritise requests for local variations in line with individual market potential.

While variation is essential, the key is to design the same level of quality into every market version. This allows you to create a strong brand that will be recognised and accepted in all territories. Marketing teams can then use that global brand strength to improve the effectiveness of their local campaigns. Although a strong global brand is helpful, you must determine which brand characteristics apply in different markets and which need to be localised. Research data suggests that brands that compete in more countries tend to have weaker scores for 'bonding' with the local market. A business model that caters to local cultures is crucial. Global product design must therefore take account of those differences. To ensure they meet local needs, multinational companies are increasingly creating global product design teams with employees or business partners from different territories.

Looking ahead, the opportunity to reach global markets is open to companies of all sizes as networked communications and Internet tools reduce barriers to market entry. Although global market entry has become easier, the challenge remains – to meet local requirements with an effective global product design strategy.

Summary

Product development is a team effort. Marketing works with product specialists, research and development groups and external partners to develop products focused on customer needs. Building on a product platform and producing different versions for specific market sectors helps you to achieve the differentiation that is essential to marketing success.

4

Offer a Customer-Focused Service Strategy

Winning technology companies recognise the importance of optimising the entire journey of customer experiences from initial enquiry to the end of product life and design their business to deliver a superior overall experience at every stage. This attitude overcomes the problems of commoditisation of products and recognises the critical role that non-product attributes play in technology markets. The key to success is to understand the entire customer journey from the customers' perspective and take the opportunity to deliver added value with professional services. The customer journey progresses through the following stages: planning, research, presales support, solution design, purchase, delivery, installation/configuration, training, management, maintenance and support, upgrade and disposal.

By working in partnership and delivering services, you gain an understanding of the problems and costs customers incur in implementing products. They may have to acquire new skills to install the product, set up training programmes for users and support staff, and provide a high level of support during the implementation period. To help customers handle implementation more efficiently and cost effectively, you can offer services such as carrying out initial installation, providing the services of an engineer on site to improve the skills of customer support staff and provide any specialist support needed. This represents value for money because the customer does not have to recruit additional, costly staff

Pricing can be improved by offering customers a package of products and services that represent real value for money for their business. A manufacturer who includes installation, training and initial support in the package offers better value for money than a competitor who simply sells the product to customers and expects them to meet the costs of start-up. Products and services can be differentiated on the basis of their value to the customer. Commodity products and services have little visible difference – price is similar, quality and performance equivalent, and the results similar. However, if a service is directly targeted at business needs, it provides a clear differentiator. Examples include:

- Training services to ensure customer staff get the best results from their systems;
- Project services to help customers complete essential projects faster and allow their own staff to concentrate on key business activities;
- Contract design and research services that customers can use to push their own product development programmes ahead at a faster rate and keep ahead of their competitors.

Supporting Business

Services play a valuable role when they support a customer's core business. The service then becomes essential to the customer's own business. For example, if a company wanted to introduce a telemarketing programme to support its own customer relationship programmes, but didn't have the skills, facilities or resources, the supplier could provide the telesales infrastructure and integrate it with the customer's own marketing activities. The customer could then concentrate on its core business development, without diverting scarce resources to handle the support activities.

Speed is an essential element in markets where the pace of change puts increasing pressure on rapid product development. If your company offers new product consultancy and provides a range of support services, that will enable the customer to achieve faster product development. Customers can get new products to market or introduce new services more quickly without committing their own resources. Getting there more quickly means that they can stay ahead of the competition and improve their response to changing market conditions – both essential to maintaining long term turnover and profitability.

The pressure of day-to-day activities prevents many companies from concentrating on new product development or important business developments. The result is that they lag behind the competition and miss the opportunity to succeed through innovation. As a strategic service provider, you can support them, either by handling the day-to-day tasks on a cost-effective basis to free the internal team or by taking on strategic development work in collaboration with the customer. Either way, it will speed up the process and provide the customer with a faster return on investment.

Sometimes companies are prevented from exploiting profitable new developments because they lack the skills or resources needed for success. By filling that gap in the customer's resources and demonstrating how it can be used to provide a strategy for growth, you can position your services as strategically important.

Customers also benefit from a reduction in the cost of their business processes. Using their own staff to handle everything can be expensive because certain skills may be used only occasionally. By analysing all their activities and dividing them into core and non-core activities, a company can decide where support will be beneficial. The support services they buy in are strategically important. A specialist service provider has the skilled staff and the infrastructure to deliver the service cost effectively and reduce the overall cost of that business process.

If your company has a reputation for supplying high-quality but expensive products, you can justify the cost difference by offering added value through a package of strategically important services. Major expenditure can be justified by its return on investment – how well and how quickly the expenditure pays for itself in benefits to the business. For example, a major consultancy project which improves a customer's competitive edge may be expensive, but it will pay for itself in enhanced revenue, profitability and market share, providing a high return on investment. Strategically important products and services are more likely to provide a good return on investment than commodities.

Recognition of the strategic importance of services has accelerated growth in the service industry as companies recognise the benefits of outsourcing activities that are not part of their core business. Companies also recognise how professional services such as consultancy, training or marketing can help them improve their business performance. Outsourcing those services allows the customer to concentrate on the activities that are critical to business success. Services can help a business to improve efficiency and achieve strategic business objectives. For example, consultancy services can help a company focus on critical areas for growth. Training services are helping companies develop their workforce skills as a competitive advantage.

Facilities management is another strategically important service in the technology sector. This sample marketing plan outlines the issues that companies face in marketing the service.

Companies utilise facilities management to operate, maintain and improve their infrastructure. They can use internal resources to implement the plan or outsource the service to a specialist facilities management company. A facilities management marketing plan must operate at two levels: it must convince potential customers of the benefits of outsourcing, and it must demonstrate that the outsourcing company offers the best solution. The main reasons why companies outsource facilities management include controlling costs, freeing capital funds, improving return on investment, and reducing labour turnover and training costs. The sales plan for service providers selling to prospects not currently outsourcing must incorporate a presentation to explain and demonstrate the strategic benefits of outsourcing.

Despite the potential financial and operational benefits of outsourced facilities management, companies have concerns about outsourcing. The two major issues are loss of control over strategically important assets and quality of service. As a result, some have brought facilities management back in-house to regain control and improve costs or response times. The marketing plan must show how you will address those concerns and communicate messages about partnership and quality of service.

The plan should indicate the types of prospects to target. A service provider can segment the market in a number of ways: geographically, concentrating on specific areas; by industry sector, such as utilities, government organisations or manufacturers; and by number of locations, including companies with single or multiple locations.

The choice of target sector depends on your facilities management capability. The plan should indicate how your capabilities correspond with the needs of different market sectors. The capability section of the plan should cover factors such as staff numbers and skills, sector-specific experience, industry certifications and approvals, quality standards, service level agreements and customer service standards.

The plan should outline the stages of the sales cycle and indicate how other marketing activities can support the sales effort. Advertising and direct marketing help to create awareness of your capability and generate leads for the sales team to follow up. Presentations and marketing material such as brochures and case studies help the sales team to present company capability to prospects. During the period when prospects are evaluating potential suppliers, regular communications by email help to maintain awareness and build preference.

Services and Brands

Service products have a direct impact on the strength of a brand because they influence a customer's experience of the brand. Quality service improves customer perception of the brand; poor service has the opposite effect. One customer's poor experience can have a multiplying effect as customers share their experience through social media. Brand strength is determined by two main factors – customers' satisfaction with the product itself and their experience in acquiring or using the product. Service products create an interaction between the customer and the brand that must be managed carefully. If a customer chooses a brand and orders it by telephone, the quality of telephone response and call handling affects brand perception. If the customer signs a maintenance contract for the product and service delivery is poor, that experience also impacts the brand.

Service products create relationships between a brand and a customer. In consumer markets the service relationship might be simple with the customer taking out a finance agreement and an extended maintenance warranty. In business-to-business markets longer-term service relationships are more common. Here, service products might include consultancy, installation, training, maintenance and upgrades. Each of those service relationships creates a customer experience and an opportunity to strengthen – or damage – the brand. The brands that are transforming markets are doing so because they see everything they do as part of the customer's experience. That means companies must set and maintain quality standards for every service product and every point of customer contact. To build and strengthen a brand, the investment in service quality is as important as the investment in product quality.

The growth of social media is encouraging consumers and business customers to share their views and concerns about products and services with a consequent impact on brand perceptions – a phenomenon known as voice of the customer. Voice of the customer programmes will increase in importance as more brands start to recognise the financial impact of poor customer service. The programmes monitor feedback in social media and manage companies' responses so that they can regain control over brand perceptions.

Service products have a double impact on brands. They can strengthen the intangible value of the brand by increasing and improving customer experience of the brand. Service products can also increase the financial value of the brand by generating additional revenue streams. That makes service development and quality control a key element in brand management.

Service Revenue

A systems company used to include services like installation, maintenance and support as part of the overall purchase price as part of an incentive to buy – a decision that was popular with the sales force. When research indicated that customers found these services valuable and would be willing to pay for them, sometimes from third-party suppliers, the company set up a service division, developed a portfolio of chargeable services and launched them. Within a year, that division became the most profitable in the company and the sales force suffered no loss of business.

That example demonstrates the value of customer services. Service is frequently relegated to maintenance and problem-solving. However, it can be a key differentiator between you and the competition. Meeting customer requirements in the most appropriate and efficient way adds enormously to the perceived value of your product. Services can add further value to a product, providing incremental income and increasing customer loyalty. Services provide you with an opportunity to continue dealing with a customer long after the initial sale. Many companies have internal service departments. They can be expensive to maintain, however, and are sometimes lacking in essential skills. By demonstrating the potential savings and benefits of outsourcing service, you can persuade them to switch to an external source.

Many companies have recognised the importance of service to their customers, and have changed their service strategy accordingly. Instead of offering free service, they have upgraded the services, widened their portfolio and started charging customers for services. Although customers may initially object to being charged for something that was free, charging demonstrates the value of the service. To take full advantage of the service opportunity, it is important to explain the benefits of effective service to customers, and present your service operations as convenient, cost-effective, and strategically important.

Customers require many different services during the time they own a product – the ownership cycle. Their requirements could include advice, consultancy and design before the sale, installation and training, followed by maintenance, upgrading and other after-sales services. Each of these represents an opportunity to earn incremental income and maintain contact with the customer. Outsourcing is a growing trend in companies seeking to concentrate on their core business, rather than attempting to do everything themselves. By using outside specialists to handle activities such as maintenance, systems management and customer support, your customers can make better use of their own resources and benefit from a quality, cost-effective service from suppliers who specialise in the activity. Outsourcing provides you with an opportunity to build closer relations with your customers by offering them vital services.

Account Control

Basic services such as installation, maintenance and upgrades are available from many different service organisations. They do not differentiate you and they do not add value. Higher-value services, requiring skill, knowledge or experience, are the keys to success. To add value to products and to increase customer loyalty, put together 'bundles' of products and services that reflect customer needs. Look at your customers' buying cycle. How frequently do they make purchases – monthly, annually, every three years, every five years? The longer the purchase cycle, the more difficult it is to retain account control. Other companies can be talking to your customers, users may be experiencing problems that you are not aware of, and the decision-making team may be changing in ways that you cannot influence. Loss of contact could mean loss of control. In that situation, services can play a key role in maintaining contact and retaining account control.

This is particularly important in technology markets where new product sales have similar purchase lead times of several years. Computer manufacturers had been losing maintenance business to independent service companies;

they also found that the customers' information systems strategy and choice of systems was being driven, not by the manufacturers, but by management consultancies. When management consultancies moved into other areas of information systems service such as application development and managed service and, when independent service companies expanded their activities, the computer manufacturers lost even more account control. Although service companies and management consultancies were not the manufacturers' direct competitors, they were enjoying high levels of contact with key decision-makers at senior and middle management level, and this influenced future business opportunities. By introducing a broad range of services, the manufacturers would be able to build high levels of contact with decision-makers throughout the customer organisation.

Customer contact is one of the most important benefits of a service programme. Here's what happened when a computer manufacturer decided to widen its service portfolio. It was able to identify a number of stages where customer service can be used to increase the frequency and quality of customer contact.

Presales consultancy – helping the customer develop a strategy in line with corporate objectives. This brings the company into contact with the senior executive team and provides valuable information on the customer's future business plans.

Planning services – turning the strategy into a practical solution. This provides contact opportunities at senior executive and operational levels.

Implementation services – helping the customer to install and introduce a new product without recruiting specialist staff or overloading their own support staff. This helps to ensure that the customer's product is implemented effectively, increasing customer satisfaction and loyalty.

Training services – providing the customer with skills development and building useful contact with departmental managers and users.

Managed services – which cover a wide range of maintenance services. This provides the supplier with one of the best opportunities for continuous contact with the customer and gives the supplier a valuable insight into the customer's changing product needs.

'Through-life cost' is a concept used in engineering to describe products that have a value beyond their initial purchase price. For example, a component that is higher priced may be simple to assemble, easy to maintain and have a long life. The costs of that product over a whole life are likely to be lower than those of an inferior product that may cost less initially, but prove expensive in the long run when installation, servicing and replacement costs are taken into account. By working with the customer to identify and reduce through-life costs, a company can increase its account control.

Another key concept in the customer relationship is the life cycle. By providing a range of services to supply the skills and resources customers need to complete a project and continue delivering the highest standards of service to their own customers, the supplier becomes involved at every stage in the life cycle of customers' products. This might take the form of helping customers reduce costs, speed up response times, improve product quality or supplement internal resources – valuable services that help to build a dependent relationship.

An example is a strategic consultancy service aligned to the customer's future business objectives. The supplier can help the customer identify the key growth paths and the skills needed to achieve those objectives. They can prepare a project plan and work with the customer's project team to meet the key requirements in a cost-effective way. They can also provide a range of services, enabling the customer to compete effectively at every stage in the life cycle.

Service Opportunities

One of the most effective ways to identify service opportunities is to look at the problems your customers face by analysing a series of business scenarios. Some examples show how the process works.

Objective advice – Your customers need to ensure that they have devised the right strategy to meet their business objectives. They need objective advice and guidance to improve the quality of their own decision-making. You could meet those requirements by offering consultancy services.

Defining problems – Your customers have identified certain activities which are crucial to their business success. They need help in defining the problems and planning the most appropriate course of action. Consultancy will also be relevant here.

Adapt quickly to change – Your customers need to adapt quickly to changing market conditions or competitive threats, but they do not have the resources or skills to succeed. You can offer your customers your skills and resources on a project basis so that they can overcome short- and long-term requirements.

Develop new skills – Your customers need to develop new user and management skills so that they can get the best return from the products and services they have bought from you. You can offer your customers training services.

Continued service operation – Your customers need to ensure that their products are continually operational and providing the business benefits they were designed for. By offering your customers maintenance services or managing their equipment for them, you can ensure that their products are kept in the best possible condition.

These business scenarios help you to identify opportunities for offering services to your customers. Customers may provide many of these services from their own resources, but it is possible to get further involved through the growth in outsourcing.

Service Standards

These are some of the key features that customers are looking for in a service offer:

- One contact point to simplify contact and service administration;
- Nationwide resources to provide support available wherever the customer is located;
- Direct contact with a technical specialist for immediate response to problems or queries;
- Quality support to recognised standards providing independent re-assurance that service standards are high;

- Complete support solutions to meet all a customer's needs with consistent service standards;
- Service options offering a choice of service levels which can be aligned to customer's needs;
- Investment in support demonstrating long-term commitment to the customer

Whether your customers have a technical query, a service request, or a product enquiry, or need advice, guidance, or information, they should be able to call one number for direct access to all your support resources. Ideally, you'll have specialists on the spot to deal with your requests. If they can't answer the query straight away, make sure that the right person calls the customer back.

Your customers may have a technical query and want to talk to an experienced specialist straight away. When they call the technical help desk, they should be talking to a highly skilled person with extensive technical support and field experience.

Superior service standards can differentiate a supplier, so it is important that you position your company as a quality service supplier. Your customers will assess your performance according to a number of factors:

- Adherence to an independent quality standard;
- Your service response mechanism – how quickly and easily can customers contact you when they have a query or a problem;
- Your service infrastructure – spares level, service network, location and number of service specialists, investment in service tools;
- Your service performance – response times, success rate, your measurement of customer satisfaction and your escalation procedures for problems that cannot be resolved.

You should be able to provide complete support throughout the time your customers own the product. Develop a comprehensive service portfolio including:

- Consultancy;
- Application development;
- Finance;
- Installation;
- Systems integration;
- Training;
- Planned maintenance;
- Technical support.

Provide your customers with a choice of flexible service options to suit their operational needs. If your customers have an in-house support operation, support their team with an efficient spares delivery service or manage their spares for them. Also offer to enhance the skills of your customer's in-house team with training, advice and guidance, technical support and access to specialists.

To meet those requirements and deliver quality service requires a significant investment in the service infrastructure. The right service team, efficient service communications and a sophisticated service management system enable you to enhance your response and performance even further. Customers expect a quality service. If you fail to deliver the right standard of service, you could damage customer relationships. That means you have to invest in people and infrastructure. You can either build your own service team through recruitment and training or work in partnership with a specialist organisation which will deliver service on your behalf.

The IT help desk is the first point of contact for IT users who have problems or need help. An effective help desk can help to maintain IT users' productivity and improve customers' Return on Investment in IT. However, Industry surveys indicate low levels of satisfaction with support from help desks. Following best practice is essential to improving the level of service to users. ITIL, the Information Technology Infrastructure Library, publishes a series of best practice guides that are recognised as an international standard for the

IT industry. The library includes a guide to help desk best practice covering the key functions of incident management, problem management, change management, configuration management and release management.

The help desk team can speed up support by publishing frequently asked questions on a website. The questions should cover only the most common issues and customers should be able to resolve the issues without the need for special skills. The help desk team can also offer training to users covering the same types of issues. Encouraging self-help reduces the workload on the help desk, giving the team more time to concentrate on urgent problems. An effective ticketing system enables the help desk to prioritise support requests and manage the most urgent or potentially damaging incidents ahead of routine issues. The system should also have the capability to route requests to team members with the appropriate product knowledge or skill level. Prioritising support requests makes the best use of support resources and can also help to improve customer satisfaction levels.

IT issues create inconvenience for customers, so it is important to communicate throughout the support process. Let customers know the target time for response and resolution and update them on progress. If the team cannot resolve an incident within the service level agreement period, operate an agreed escalation process. For upgrades or other systems changes, let customers know the planned programme times to minimise inconvenience.

Help desk software should have the facilities to capture and analyse incidents to build up a picture of recurring events. This will make it easier for help desk teams to plan and implement problem management. Setting up a community forum on your website can also help to capture and share knowledge on recurring issues. Choosing the right metrics is essential to monitor performance and plan future resource requirements. Metrics that simply record the number of requests and time to respond may not provide a complete picture of help desk effectiveness. Measuring users' satisfaction with the level of service provides a basis for planning improvement programmes.

Summary

Services play an important role in technology marketing. They enable you to deliver a complete solution to customers with services that support key business processes. They also help to build relationships with customers throughout their ownership of a technology product. Services increase revenue, help to differentiate products and contribute to increased customer satisfaction.

5

Operate a Flexible Pricing Strategy

Customers buy products on the basis of the price or the value the product offers. If customers believe that your product offers a higher level of value than your competitors, they may be willing to pay a higher price to obtain the benefits. When your products do not offer any significant value to the customer, price becomes the more important consideration. Your sales representatives must therefore understand the aspects of a product that represent value to customers and communicate those benefits. Lack of value forces sales representatives to sell on the basis of price alone.

Value is not an inherent feature of a product. Customers recognise value when the product helps them to solve a problem or improve their business performance. A product offers value when it helps a customer reduce costs, improve productivity or quality, move into new markets or eliminate waste, for example. Intangible benefits such as speedy delivery, responsive customer service or long-term guarantees also add value to a product.

Managing Value

The combination of information on the Internet and the growing use of feedback and reviews on social networking sites enable customers to gain a greater understanding of the range of products and services available to them. That means customers now have an improved ability to articulate their value expectations. Knowing what customers value and how the solutions your company provides meet those values is key to developing a viable marketing strategy.

By monitoring customer feedback and reviews, you can identify the factors that customers value most. Those factors form the basis for value-based pricing. The change in pricing strategy enables you to focus on factors that differentiate products from competitors and calculate the price premium they can charge. By understanding the factors that customers value, you can position your products and communicate messages that underpin the premium pricing. A product that reduces a customer's maintenance costs demonstrates value over the lifetime of a product. Services that help customers run their business more efficiently or develop new products demonstrate strategic value. Both factors are strong differentiators and justify premium pricing based on customer value.

Product development strategy should also be based on customer value – what improvements to your product would add the most value from the customer's perspective? As well as generic product developments, companies should also consider customising products for individual customers where there is an opportunity to add specific value. Factors based on value enable your sales representatives to differentiate a product from competitors. Without the differentiators, it can be difficult for sales representatives to present a convincing case for selecting your product over a competitor's equivalent. If your sales representatives do not have the skills to sell on the basis of value, you should consider retraining the sales team or recruiting sales representatives with experience in selling on value.

When a group of products from different suppliers have no differentiating benefits that represent value, the products become commodities. Price then becomes the only distinguishing characteristic. If your sales representatives have to reduce prices to win or retain business, the result will be lower income and reduced profit margins which could prove damaging to your business.

Lack of value also makes it difficult to build relationships with different decision-makers in a customer company. When companies buy commodity products on price alone, they leave the buying decision to a purchasing manager. Selling products on the basis of value brings other decision-makers into the purchasing process. The greater the value you offer, the more people will influence the buying decision. For example, a product that enables a customer to move into profitable new markets will be of interest to senior decision-makers including chief executive, marketing director and finance director.

Customer Dependency

It's important to understand how important your product or service is to your customers. If your customers are dependent on your product or service, you can base your pricing on that factor, rather than your own costs or competitors' prices. This is called customer value pricing. However, the basis for customer value can change in times of recession. When economic conditions are good, customers put a higher value on products and services that give them a strong competitive advantage, for example by improving their quality or increasing their production capacity. In a recession, the emphasis may be on products or services that ensure survival, for example by reducing their costs.

The more you understand about the problems your product solves for the customer, the easier it is to tailor your offer and get the best possible price for each sale. Although you may believe that your products are important or even essential to customers, the market may have a different perception of value. Experience indicates that, within a single marketplace, some prospects will feel that products are too expensive, while others would have continued to buy, even if prices were higher. Interview your customers to find out how they view your products and services. By building a better understanding of market perceptions, you can segment and target your prices to meet different customer requirements. Any knowledge of the value you deliver to your customers gives you greater control and confidence in your pricing.

The segmentation approach allows you to offer different customer groups a range of prices or price packages for the same basic product. For example, you could offer 'first-class' or 'economy' prices. First-class customers receive extra value, such as faster delivery or complementary services by paying a premium price. Economy customers pay a lower price but would get just the basic product, with no added value. This type of price segmentation can help you reach different markets without changing the specification or costs of your basic product.

A good pricing strategy recognises the importance of protecting strong brands. You should keep high-value products priced appropriately. Using price cuts to maintain business on high-value products or eroding margins by giving away services or offering discounts to customers can prove a false economy. That can erode the base of profitable customers and reduce the potential for profitability when the downturn ends.

While it may be important to offer a competitive headline price, you should also consider opportunities to maintain profitability by recovering some of the costs involved in order processing, delivery, financing or other customer-facing processes:

- Set minimum order quantities to minimise order processing costs.
- Focus on collection efforts to improve cashflow.
- Offer customers a priced menu of essential services such as delivery or preferential discounts.
- Identify opportunities for incremental sales of value-added services such as consultancy, planning, installation, training and maintenance.
- Offer preferential discounts for long-term contracts to smooth out revenue.

Life Cycle Pricing

Another approach to offering value is to base pricing on life cycle costs. This enables you to calculate the total cost of owning and operating a product from initial purchase to the end of its working life. The costs include purchase price, financing, installation, training, operation, support, maintenance and disposal of the product at the end of its life. This is called the total cost of ownership. You can use life cycle cost estimates as part of a marketing programme to demonstrate to customers that a product has a lower total cost of ownership than a competitive product with a lower selling price.

Life cycle costs vary by type of product and pattern of use. A customer planning to buy a sophisticated network solution will consider a number of life cycle costs. The network is complex and will require installation by skilled service staff. The company will have to recruit and train skilled operators who will command a high salary. To avoid downtime and disruption to production, the company will have to take out a maintenance contract that guarantees rapid response times to service requests. Life cycle costs for that type of solution are high.

A company that buys products that it will only use occasionally will have different requirements for life cycle costs. Maintenance costs, for example, would be less important. Instead, the company may want to sell the asset into the second user market at the end of its working life. It would put a high resale value as a priority in analysing life cycle costs.

Life cycle costs can be difficult to estimate if external factors affect them. Equipment with high energy requirements, for example, can incur increased operating costs if fuel prices rise significantly. Wage inflation can also impact operating costs, particularly if the equipment requires skilled operators.

You can reduce your customers' life cycle costs by focusing on areas of high cost. Improving the reliability of components or using higher quality materials can reduce maintenance requirements and associated costs. Making the equipment simpler to operate or introducing higher levels of automation can reduce the need for highly paid skilled operators. Increasing fuel efficiency can reduce the impact of future energy cost increases. Design changes like this can have a positive effect on maintaining or reducing life cycle costs, making your products more competitive.

A network company used its technological advantage over competitors to reduce customers' life cycle costs. The company's microwave technology has a 'Mean Time between Failure' (MTBF) rate which was the best in the industry. This meant that the equipment failed far fewer times than competitors' equipment, resulting in fewer callout charges and lower logistics and spares costs, resulting in both lower Operating Expenditure (OPEX) and lower Capital Expenditure (CAPEX). The company was able to demonstrate a lower total cost of ownership, despite its higher initial price.

Outsourcing can also contribute to lower total costs. The effects of the recession on staff reductions and budget cuts have been a contributing factor. IT managers can choose to outsource a wide range of services from application development through data centre operations to IT security, depending on their own internal cost structures. Lack of in-house skills can also create a demand for outsourcing, particularly for projects involving emerging technologies or increasing complexity. IT security, where threats are growing increasingly sophisticated, offers considerable potential for improvements in service levels through outsourcing.

Although offshore outsourcing has continued to grow, IT managers now have a wider range of options for supplementing internal resources and reducing costs. Infrastructure as a Service provides a company with additional computing resources, with no need for investment in new infrastructure. Instead companies acquire the cloud resources on a pay-as-you-go basis, allowing them to scale their infrastructure immediately and cost effectively. Platform as a Service is another cloud service that allows organisations to scale their development facilities for the duration of a project. This is an emerging alternative to outsourcing new service development and it could give you a strong cost advantage.

Recession Pricing

The early stages of the recent global recession saw a period of panic price cutting and permanent sales as businesses tried to maintain cashflow. Heavy discounting accelerated when a number of high-profile retail chains failed, although the recession alone may not have been the main cause of failure. Before the recession, some retail chains had used permanent sales as a feature of their business. Although this may have proved a competitive advantage in normal trading conditions, it may be less of a differentiator in recession.

In a downturn, price strategy should have two main aims – to maintain cashflow and to protect market share and customer loyalty. Organisations can use differential pricing and discounting to reward and retain loyal or high-value customers. Loyalty pricing can also be tied to longer-term contracts for services or product supply. This kind of pricing would be appropriate for business-to-business markets where higher-value customer groups can be easily identified. Unlike permanent discounting, a selective pricing strategy rewards loyal customers, protecting market share and strengthening key relationships.

Reducing prices to generate more sales will not improve your business in the long-term. However, cutting costs can help to ensure survival and levels of profitability. Controlling costs and streamlining processes by reducing inefficiencies allows you to offer prices at levels your competitors cannot sustain. One way to improve efficiency and reduce costs without investment is to partner with other companies in a business ecosystem. Members of a business ecosystem can utilise their partners' skills and resources for a wide range of business functions, including sales, marketing, manufacturing, technical support and customer training. This provides a flexible, scalable business model that allows companies to adapt their operations and lower costs in line with customer demand by using partner resources, rather than making slower, longer-term investments.

To prevent a price strategy from damaging your business, it's important to build profit contribution goals, as well as sales goals for products, market segments and individual customers. Setting profitability goals may take precedence over market-share goals if there is a risk to the long-term viability of your business.

In a recession, revenue losses can occur because the sales force or retail network finds it difficult to resist customer pressure for price cuts. Sales teams may find themselves lowering prices to ensure they win the sale – at any cost. To discourage or prevent this, it is important to impose business rules that relate price negotiations to different volume, delivery, financing and settlement criteria. The business rules should also reflect the value of the product or service to the customer. To reinforce the business rules, it may be necessary to change the basis of sales force compensation to reflect the contribution to profitability and customer retention, rather than pure sales volume.

Although an effective pricing strategy can help to maintain revenue and profitability, it's also important to continue investment in marketing. Typically in a recession, marketing budgets are cut in an effort to boost quarterly performance. However, with reducing demand and increasing competition, this could prove counter-productive. Industry research indicates that maintaining or increasing marketing during a recession can strengthen a company's position in the short to medium term, and in the longer term as the economy picks up.

Companies that continue to invest in marketing gain a number of benefits:

- If competitors reduce their marketing, there is less 'noise' to compete with and campaigns will gain a louder, clearer voice.
- Customers respond to suppliers who are still actively engaging with them, and delivering messages that are relevant to their changing needs.
- If brand awareness falls, market share is also likely to go down and that may be hard to win back.

- When customers spend less, every sale will be harder to win. Companies who continue to invest in marketing bring stronger resources to the fight.

The right pricing strategies will not only strengthen business in a downturn, it will also position for growth later in normal trading conditions. Companies with a strong brand image ensure that a 'survival' pricing strategy will not cheapen the long-term value of their product. They recognise that customers will not automatically accept a return to higher prices in the future, but they also realise that customers will respond to a value message that is directly relevant to them.

If sales and profits are falling and customers are demanding better deals, it's easy to respond by just cutting prices. This can boost sales quickly, helps cover fixed costs and gives the organisation 'breathing space' until trading conditions improve. However, short-term price cuts may affect profitability when the upturn occurs. It may also encourage customers to push for further discounting.

Summary

Despite the pressure on prices due to recent economic conditions, your pricing strategy should be based on the value it offers customers. Products or services that reduce customers' costs or solve strategic business problems provide an opportunity for premium pricing.

6

Broaden Capability Through Collaboration

Partnership is a strategy that technology companies use to extend their product or service capability and offer customers a more complete solution. Technology partnerships take a number of forms including alliances for sales, consultancy, training, industry specialisation, support, application development and integration. Technology companies with a very wide range of partnerships create an ecosystem of support for their customers. The relationship between technology partners features varying levels of collaboration, community, joint marketing and integration of business systems.

Partnership is essential to the successful delivery of large complex IT projects, such as the installation of a manufacturing control system. The project would cover a number of stages including initial consultancy, system design and development, application development, installation, training and ongoing support. The lead company in the project might be a software or hardware manufacturer that would partner with other companies to provide the customer with a single point of contact for the complete project.

Partnership also gives technology companies the scale to tackle larger, global business opportunities. A strategic partnership between AT&T and IBM involved collaboration on more than 500 corporate accounts in 48 countries. The companies developed a complementary product portfolio, joint forms of corporate governance and standard customer-facing tools to ensure integrated service management.

Technology companies use partnerships to strengthen their position in the marketplace. A manufacturer of computer operating systems, for example, would form partnerships with application developers to offer the market a broad range of compatible applications, extending the value of the operating systems. The developers benefit from access to a wider market, as well as high levels of technical support from the manufacturer.

Channel partnerships are an important feature of the technology sector. Channel partners distribute a company's products and services to smaller customers. They may also provide various levels of customer service such as installation, systems integration and maintenance. Technology companies provide their channel partners with a wide range of support to develop their business. The IBM PartnerWorld programme, for example, offers marketing, selling, technical and training support as well as collaboration on complex projects. The programme aims to promote partners' offerings, increase their visibility in the market and deepen the skills and consultative capabilities that are essential to market success.

Collaborative Product Development

Collaboration with supply chain partners can give your company access to new products, new technologies, new skills and new markets. Collaborating on product development programmes, for example, can help you reduce production costs, make better use of skilled, specialist resources and speed up time to market. Partnership brings together the objective knowledge, experience and specialist skills of people from different companies who can contribute to the development of your new product more effectively than individual team members in one company performing their own tasks.

This can be important if you face an increasingly competitive marketplace or move into new markets. You may be feeling the pressure of a shorter product lifecycle, as customers demand more and more innovative features and benefits. Competitive pressures mean you have to reduce the time and effort involved in bringing products to market. By collaborating and creating efficient joint project teams, you can tackle key challenges in the product development process, improve the success rate of new products and grow market share and revenue.

Collaborative product development can bring your company a number of important benefits:

- Access to scarce skills not available in your own organisation;

- Additional resources to tackle larger development projects or reduce lead times;
- Access to technologies, components or supplies that can enhance your own product.

Collaborative product development can help you get a new product to market in the shortest possible time to gain competitive advantage. To achieve this, it's vital that you and your partners integrate all product development activities within a common framework. In collaborative development, the programme can be divided into smaller critical activities which are then handled by appropriate partners. As well as design and manufacturing processes, you can also utilise external partnership services to handle market and customer research, distribution, marketing communications and product launch. By establishing long-term partnerships with specialists, you can reduce the learning period and ensure that every aspect of new product development and launch is handled quickly and efficiently.

You can also create a stronger culture of innovation by bringing in partners. Many companies are finding that, to maintain the pace of innovation and compete effectively, they need to involve external specialists in this way. Partnership allows your company to focus on its core strengths and introduce expertise, experience and skills that you may not have inside the company. That means including suppliers, partners and customers in an extended enterprise with the aim of creating higher-quality products through increased innovation and a fresh perspective that can stimulate radical new solutions.

In the biotechnology sector, partnerships with universities, specialist research laboratories and health organisations enable manufacturers to accelerate their own new product development programmes and ensure that they are at the cutting edge of industry developments. This has encouraged many companies in this sector to locate in areas that are regarded as centres of excellence in medical research, such as Cambridge or Oxford in the UK.

Those external technical skills and resources can also help you increase the level of innovation without investment in your own skills. You could use partners' technical resources to handle product development on a subcontract basis, gaining access to specialist resources or additional research and development capacity to improve the performance of your own product development programmes. Alternatively, you could use partner's technical expertise to develop new products that you could not achieve yourself. This provides you with new technology and allows you to diversify in line with their specialist skills.

As well as improving your own product development process, you can use partners' products and services to expand your own portfolio and increase potential revenue and account control. For example, by using a partner's service organisation, you can offer your customers a wide range of after-sales support services without setting up your own infrastructure. This provides new sources of revenue and also increases your control over the customer.

An IT services organisation that began life as an internal company division, supporting only the company's own products recognised that, to succeed in changing market conditions, it would have to support products from many different companies. It did not have the skills to support other products to the same quality standards as its own products, so it set up alliances with other service organisations. However, to maintain quality standards, the company set a series of service level agreements for partners and backed those with training in customer care. As a result, the services organisation was able to win a much wider variety of contracts and build its reputation in new markets.

Partners' products can fill gaps in your own range or provide products that enable you to enter new market sectors. You can treat the partner products as 'white label', market them under your own brand or retain the partner name and draw on the strengths of their marketing. As well as filling in gaps in your own range, you can use partner products to offer customers higher- or lower-priced versions of your existing products, expanding your market opportunities without compromising your core product values and quality.

Geographical Collaboration

If you are considering expansion into regional or international markets, partnership can provide you with a 'shortcut' to local representation. If your partners have a network of local branch offices or distributors, this can provide you with a ready-made infrastructure for providing a local service to your customers. This approach has a number of potential benefits:

- Partners' local knowledge and contacts that can help you establish your own network;

- Existing branches that you can use to distribute your products;
- An opportunity to work on a network basis, where the partners operate in geographically separate territories but cooperate and share facilities so that each partner can offer its customers a broader service.

A software distributor that wanted to expand its international presence recognised that effective communications would be a vital success factor. The company was able to set up a joint venture with other established distributors in a number of territories, but found that the partners' existing communications networks were inadequate. To bring communications up to standard, it formed a further partnership with a global telecommunications company that was able to offer the company access to high-quality networking in all target territories. This gave the company the infrastructure it needed to provide customers with a high-quality service, without investing any of its own capital in building new networks.

To assess the opportunities for geographical collaboration, you need to evaluate the strength and performance of partners' networks:

- Can they provide a quality local service to your customers?
- Are their existing local networks performing well?
- How does their geographical coverage compare with your targets?
- Do their performance standards match your requirements?

Using partner resources in this way allows you to quickly establish a local presence and test the market before committing further resources to international expansion. It means you can respond quickly to local opportunities and build your international presence step-by-step without compromising existing markets.

Levels of Collaboration

Partnership can operate at many different levels in a supply chain. Supply chains are interdependent and their value is cumulative, dependent on the final customer. There are five types of relationship in a supply chain: arm's length, collaboration, partnership, joint venture and vertical integration.

In a basic supply chain, there are a series of buyer and supplier relationships with no real collaboration. This makes it difficult to integrate planning and supply, leaving the lead manufacturer at the head of the chain vulnerable to fluctuations in supply performance and quality. In a collaborative relationship, the members use networked communications tools to share data and coordinate planning throughout the supply chain. This can reduce costs by eliminating waste and improving efficiencies.

In sophisticated sectors, collaborative supply chain relationships feature different supply tiers. Tier 1 members provide strategic products or services that are critical to the success of the supply chain. Lower tier members may operate as subcontractors to tier 1 suppliers, but their services are less critical. Relationships between tier 1 members and the lead supplier are close and carefully managed by stringent quality and contractual agreements. Streamlining business processes with tier 1 suppliers can improve efficiencies further.

Partnership between supply chain members creates a more formal, long-term relationship covered by a contractual agreement. Within the partnership members may undertake joint development projects to improve products and services or reduce costs. Partnership builds greater stability into the relationship, ensuring continuity of supply and giving the partners access to specialist skills and resources.

The partnership relationships become more formal as members undertake joint ventures. These are strategic relationships covered by formal financial and contractual agreements. They allow partners to share risks and costs and give both parties the opportunity to enter new markets, increase capacity or access specialist skills or technology. Partners can undertake joint ventures for a specific project or agree to cooperate on a long-term basis.

Partnership in the supply chain reaches its highest levels when vertical integration or a strategic merger takes place. Here a lead company takes ownership of strategic suppliers to secure continuity of supply. This can improve costs and efficiency further. It can also act as a barrier to entry for potential competitors, particularly if the arrangement covers scarce resources or skills.

Companies undertake strategic mergers with other organisations to accelerate their growth, rather than growing organically. The aim of a merger is to create an organisation that is stronger than the sum of its parts. The merged organisation is then in a better position to achieve its strategic goals. Companies use strategic mergers to achieve a number of different objectives, including gaining access to technology or products, acquiring additional customers, creating or removing barriers to entry and developing economies of scale. Growth is a key factor in strategic merger decisions. Companies recognise that growth will enable them to compete more effectively against larger competitors or reduce their costs by taking advantage of economies of scale.

A strategic merger can give you access to products or services that are not in your current range. The new products may enable you to increase revenue by offering a wider range to existing customers or meeting the requirements of new customers. Acquiring existing products also reduces your product development costs and enables you to replace older or weaker products that are not profitable. You can use strategic mergers to strengthen your supply chain. By merging with a key supplier, you can protect your source of supply and potentially reduce your costs. This is important when a supplier is the only source of an essential raw material or component. This approach also creates barriers to entry for potential competitors, strengthening your competitive position further.

Where a company has a strong marketing operation or distribution network, it can use strategic mergers to acquire additional products to sell through its sales channels. An important aspect of Cisco's strategy is to acquire companies with products that complement its own. It can then use its sales strengths to sell add-on products to existing customers.

Business Ecosystems

If your company has benefited from partnerships or strategic alliances, you could make further gains by becoming a member of a business ecosystem. Business ecosystems relate supply chain partnership to a natural ecosystem where success and survival is dependent on cooperation between all members.

Business ecosystems are loose networks of customers, manufacturers, suppliers, distributors, service providers, manufacturers of related products or services, technology providers and other support organisations that rely on the strength of the network as a whole for their own success. Member companies do not depend on their own internal capabilities for success, they can draw on the skills and resources of partners. Members aim to improve the overall health of their ecosystems by contributing to a set of common assets that other companies can use to improve their own competitive performance. This might take the form of communications tools to connect members, or knowledge and resources to speed up and simplify the development of new products by other members.

Silicon Valley is an example of a business ecosystem. Within the community is the full web of expertise necessary to develop, fund, deliver, market and support a technology solution, including software developers, product integrators, resellers, specialist subcontractors, service providers, investors and specialists in technology sales and marketing. Typically, these are the types of members needed to create a successful ecosystem.

The same principles apply in industries where clusters of manufacturers, suppliers and subcontractors are often located in close proximity. This makes collaboration easier and also cuts logistics costs by reducing journey times. These clusters often generate high levels of skills that member companies can draw on and they offer innovative start-up companies the opportunity to build their business quickly. In mid Essex in the UK, a number of leading electronics companies positively encourage the development of new businesses through an innovation ecosystem that benefits all parties. However, a single geographical location is not essential; collaboration over the Internet means that business ecosystems can be built on global networks spanning several time zones.

Many business ecosystems include members who would normally be considered competitors. However, within the ecosystem, they cooperate to protect and develop the ecosystem – a situation called 'coopetition'. Members can cooperate in many different ways – defining technical standards, sharing best practices, sponsoring market studies or lobbying government. In the technology sector, for example, firms may jointly agree on technical standards and then develop products that compete with each other using those same standards. By coming together within networks, firms are able to cooperate dynamically and offer complex services, they can create new market opportunities, combine their knowledge, products and services, and jointly produce and offer new services and products.

Smaller companies outside a business ecosystem may have suffered lack of resources, skilled labour or affordable communications solutions. The new model emphasises networking, knowledge transfer and shared learning. Dynamic networking allows firms to build communities that share business, knowledge and infrastructures. Business ecosystems offer smaller companies the same possibility that large structured enterprises benefit from, to dynamically aggregate services and organisations, exchanging knowledge, finding partnerships and boosting market demand.

An important characteristic of business ecosystems is their ability to encourage and support innovation throughout the network. Lead companies help to create value for the rest of the ecosystem in different ways, but the first requirement usually involves the creation of a platform, an asset in the form of services, tools or technologies that offers solutions to other members. This can take the form of physical assets, like the efficient manufacturing capabilities that larger companies offer to members that don't have their own facilities, or an intellectual asset, like a software platform.

Members of a business ecosystem can leverage their partners' skills and resources for a wide range of business functions, apart from development of core products and business strategy. For example, they can use partners for sales, marketing, manufacturing, technical support and customer training. This provides a flexible, scalable business model that allows companies to rapidly expand or change their operations in line with customer demand using partner resources, rather than slower, longer-term investments. Using partners around the world also enables international expansion without setting up local operations.

One of the ways to improve quality and performance within the ecosystem is to develop and ratify standards. Certification, based on training and assessment, is an example of this. Higher-qualified members of an ecosystem are then allowed to take on more complex or higher-value projects. They can be ranked as 'Tier 1' suppliers for example. The cooperative nature of business ecosystems also encourages informal methods of improvement through knowledge transfer and the development of communities to share knowledge and experience on common technical issues.

In many market sectors, the business ecosystem has replaced traditional relationships of buyers and sellers with communities of interacting organisations that together create, deliver and consume goods and services.

These relationships should evolve in response to changes in the business environment, market conditions and the specific initiatives of individual members. Slowly changing, rigid supply chains are being replaced by more fluid, flexible structures based on alliances, partnerships and collaborations. This model offers both small and large companies opportunities to grow through cooperation and evolution.

Managing Collaboration

There are strong, clear marketing benefits in partnerships and this form of collaboration is now common in many technology sectors. You'll find that many companies list their partnerships and strategic alliances on their website as a demonstration of their extended capability. The most successful partnerships are those that are managed effectively and this has become an important strategic marketing role, with a number of companies appointing specialists to get the greatest benefit from their partners.

Managing partnerships has three key roles:

1. Setting and managing relationships;
2. Building and supporting partnership teams;
3. Managing the communications infrastructure.

When you are planning a collaborative project, it's important to choose the right partners. Your partners should add value to your own offer, or provide other benefits such as access to scarce skills, faster time to market or reduced costs. As well as tangible benefits, it's important that your partners share your own business and marketing values, and recognise mutual benefit in working with you.

This checklist can help you assess the potential of a partnership:

- Commitment to the project;
- Future direction;
- Market experience;

- Technical expertise;
- Management capability;
- Financial stability;
- Quality processes;
- Adequate resources;
- Policy of collaboration.

Although partners may have strong individual business and marketing objectives, it's essential that any collaborative development project works to a single objective agreed by all the partners. The success of the project could be damaged if each partner insists on following their own development agenda. The result may be a compromise that does not fulfil the new product's full potential. When collaborating companies have an effective working relationship and can agree on a single strategy, the result should exceed what they could achieve individually.

Effective partnership requires a culture that encourages teamwork, cooperation and collaboration. That means each team member must have clearly defined responsibilities within an agreed product development process based on sharing of information and collaboration. There are often conflicting goals in product development, so collaboration requires effective team work where team members trust and respect one another. There must be open communication and a willingness to accept input from others.

Collaboration has become easier through tools like videoconferencing and the use of secure extranets. Working together in the same location would be an ideal solution, but travel costs could prove prohibitive if the partner companies are far apart. However, secure communication systems and sophisticated collaboration tools mean that dispersed team members can work together effectively in a 'virtual enterprise'. By using collaborative technologies, partnership teams can share accurate, up-to-date digital product information across different companies as well as across different functions like design, engineering, manufacturing, marketing sales and purchasing. Members of the extended team can post suggestions, reviews and other feedback. It's also possible to set up virtual project rooms where teams from different organisations can collaborate on a project using the site database and pull in third-party resources as they need them.

The tools and technologies include:

- Email to exchange drawings, models and project information;
- Meetings held by teleconferencing and videoconferencing;
- Meetings held via the Internet;
- Project websites to create a single source of project documentation, with email alerts for updates;
- Websites to enable team members to view and comment on project documents or drawings;
- Tools for product data and product information management.

For effective collaborative development, information must be shared more frequently, in a wider variety of ways among dispersed team members. With the right technology solution, you can create an open network that supports communication and collaboration across all the parties involved in a joint product development project. Team members can access knowledge that has typically been trapped in departmental systems, and the collective knowledge allows the team to develop and go to market as one integrated company, rather than a number of separate isolated groups.

When team members are located in different sites or even different countries, it's essential to ensure that everyone is working with the same up-to-date information. When data is created at multiple sites, it can be difficult to track and maintain an up-to-date history of the communications and decisions relating to the development project. The process can slow down if essential data is missing, out of date or incorrect. Secure access to current information gives the team confidence that, as changes occur, they're precisely tracked.

Summary

Partnership and collaboration can add value to your marketing programmes, helping you extend your capability and product range, speed up time to market and expand your market opportunities. Partnership is now recognised as an integral element of the technology business, with small and large players participating in ecosystems that benefit both suppliers and customers.

7

Make Knowledge a Differentiator

Acquiring and communicating knowledge is an important aspect of technology marketing. By educating and informing customers, you can help them make more effective decisions about products, services and suppliers. You can also use knowledge to improve the way you target customers and deliver service.

Customer Awareness

Customers are changing the way they find out about products. Instead of relying solely on companies' marketing material, they use independent websites and social networking tools to research products before they buy. They study reviews and feedback from other customers which they believe provides a more objective view than company marketing material. To regain control of communication channels and maintain awareness, you must help customers when they search or exchange opinions. Setting up a forum or review page on a website and responding to customer concerns can help you raise awareness and build trust.

A marketing strategy that raises market awareness can also help you become the recognised leader in your field. When customers and prospects are fully aware of your company, its products and the product's benefits, the barriers to a sale come down. In existing sectors, increased awareness helps you sell more, launch new products successfully or increase market share. If you are aiming at new markets, raising awareness creates a platform for successful market entry.

Customers may not be fully aware of the benefits of a product or service to their business or their personal life. The more complex the product, the more difficult it can be to make customers aware of the benefits. Marketing strategy should therefore focus on customer education using communication tools such as case studies, informative articles in the press or on websites, training material and customer events such as workshops or conferences. Raising awareness of product benefits helps customers make informed purchasing decisions and can increase sales as customers gain a greater understanding of the value of the product.

When you are entering a new market, you have to create awareness of both your company and your products. Why should customers choose products from you, the new entrant, rather than existing suppliers? What credentials do you have? What is your track record in other markets? Marketing strategy has to build awareness of your company rapidly so that other marketing and sales initiatives hit their targets and capture share.

Market Education

Market education can change the way customers perceive your company with a resulting impact on sales. Customers are more likely to buy from a company that they believe is innovative, responsive and in tune with their needs. In the business-to-business sector, customers also want to know that their suppliers are well managed and financially stable so that they can rely on them to deliver quality products on time, every time. To raise awareness of your company profile, your marketing strategy should include corporate communications in the press, success stories and news about investment, innovation and financial performance.

Market education can help to meet many different aspects of your marketing strategy. These are some examples:

- Increase sales to existing customers;
- Improve sales force productivity;
- Open new revenue streams;
- Position the company in different ways;
- Ensure the success of a product launch.

Before you consider a market education programme, carry out product and market research to ensure that you understand your customers' needs. Work with them to help them develop wish lists of features they are looking for in new products. In your research, ask questions to see if customers understand the full benefits of new features that you plan to introduce.

As product development progresses, conduct further customer research, showing concepts and analysing response to identify any issues in understanding. Use this research as the basis for a market education programme to build momentum and understanding before the launch. The key to success for companies in this situation is to build trust and position themselves as advisers, rather than sales people.

Articles in important publications, speaking appearances at industry events, well-produced customer magazines, white papers, podcasts and webcasts can all help to build understanding and preference for the company's products and services and position a company as a trusted expert.

Providing quality keynote speakers at a customer event forms an important part of the customer education process, as well as raising your company's profile you can achieve the same results by getting on discussion panels or conference workshops. If you are participating in an external event, organisers publish dates for submission of conference papers so make sure you are aware of those dates and provide suitable material. Again, focus on providing useful information. You are there to inform and educate, to provide a unique perspective. If your speaker is selected, promote their presentation in pre-event publicity and publish their papers in your post-event communications.

Executive briefing meetings give you the opportunity to bring your customers up to date with new developments in your business or in your industry that might benefit them. For example, you might brief them on new technical developments or new legislation that is likely to impact on them. This type of meeting not only demonstrates your professionalism and builds thought leadership; it also helps to add value to the customer relationship. This is an important part of the process when the discussion of new products needs to move from technical to business decision-makers.

Cisco Systems recognised that communications networks were not just transport systems for voice and data traffic, they were powerful business tools capable of transforming the way companies operated and collaborated with other organisations. To communicate that message and build understanding, they had to move the discussion from the IT department to the board room with a programme of executive education.

Thought Leadership

Raising market awareness can be the first step to positioning your company as the expert or leader in a sector. By providing information that helps customers overcome their business or personal challenges, you can position yourself as a trusted adviser and differentiate your company from competitors. This strategy of thought leadership can provide an important boost to long-term revenue and profitability.

A thought leader is a company that actively promotes and discusses ideas that are relevant to their marketplace. By talking about a topic with authority, you can become recognised as the leader in that field. As a result, the market assumes that your company has the experience and knowledge behind it support what you are saying. If you become a thought leader in your field, it doesn't matter how big or how small your company is. Customers and prospects look to your company for insight and vision, journalists will quote you, analysts will call you and websites will link to you.

Most companies have to work hard to get visibility, especially in a crowded marketplace. If your reputation is not strong, you may find that prospects don't want to buy from an unknown vendor. They prefer to buy from companies they can trust with confidence. Trust is built on reputation and reputation is generally not built on advertising. By becoming an industry thought leader, you can build a strong reputation, even with limited resources. Thought leadership supports your marketing and corporate objectives to sell products, generate leads or boost share price.

Make thought leadership a strategic imperative for your company. Your company does not need to be a leader in sales to be considered a thought leader. However, becoming a thought leader does require work and commitment. It demands the often difficult task of looking at your company from the perspective of the world outside. Generating an ongoing effort towards thought leadership is the best way to ensure that it actually happens. The key elements of a thought leadership strategy include:

- Press relations;

- Customer magazines;
- Conference presentations;
- Executive briefing meetings;
- Website content;
- Podcasts;
- Webcasts;
- Blogs.

Relations with the press form an important part of a thought leadership strategy. You should identify the most important writers and editors who report on your market. As well as providing them with your normal product and corporate press information, you should keep in regular contact, offering them comments or articles on important issues in your industry. If a newspaper or trade magazine is running a supplement on your industry, offer articles and other content. By maintaining regular contact, your company can become a trusted source for journalists, helping to increase your visibility in the press and build your reputation.

You can use your customer magazine to position your company as a leader in its field. The majority of the content would be contributed by independent writers on issues of general importance to the industry. There would be minimal reference to your own products or services and any company branding would be very subtle. A customer magazine used to develop thought leadership could include different types of content such as:

- Feature articles;
- Case studies;
- Interviews with experts or senior company executives;
- Discussion panels on important topics.

Your website should provide a source of useful information for customers, prospects and influencers. Your website increases in value the more people know about it and link to it. You should therefore include:

- Copies of white papers and case studies for downloading;
- Bulletins on research you are carrying out;
- Electronic copies of your customer magazines;
- Copies of seminar or conference papers delivered by your own speakers;
- Copies of published articles or news items that demonstrate thought leadership;
- Details of events where your company is participating;
- Blogs commenting on industry issues;
- Archive copies of webcasts or podcasts for downloading.

A podcast is an important element in developing thought leadership. If you have experts on particular topics, you can create programmes giving advice and guidance or opinions on important industry topics. You can also host discussion groups that might include your own staff, customers, suppliers and industry consultants. The discussions could again focus on important industry topics. To create a more interactive programme, you could invite questions or feedback from customers which could be added to the podcast.

Webcasts, like podcasts, enable you to develop and publish thought leadership programmes such as discussion groups or advice and guidance via the Internet. Webcasts can also be designed to be interactive between the presenter and audience. Webcasts can include:

- Slide presentations;
- Live video;
- Text chat for live question and answer sessions;

- Polls and surveys.

Weblogs, also known as blogs, have become a popular channel of communication in recent years, creating thought leaders in the process. Blogs can provide excellent insight into the way a company works and provide a platform for commenting on industry issues. Commentators believe that companies are now less guarded in sharing intellectual property or thinking in a field. Blogging is therefore an accepted channel of open communication. Although blogging appears to be an informal medium, planning can be important if you are using a weblog as part of your thought leadership strategy. Where possible, you should produce a publishing calendar to ensure you add new content at regular intervals. This will encourage visitors to return to your site to look for new content.

White Papers

White papers and case studies help to build understanding and awareness. They are useful resources that demonstrate you understand and care about the problems that your customers are trying to solve. Your white paper should review problems or needs faced by the reader, rather than describe a product or service offered by your company. White papers are an important tool for communicating with technical decision-makers. They are widely used by companies marketing business-to-business and technology products. White papers help people make decisions. In the IT sector, industry research indicates that around 70 per cent of information technology professionals in the USA rely on white papers to make purchasing decisions. They are particularly valuable in the early stages of purchase when decision-makers are gathering information.

They should be clearly written and illustrated and may need the skills of a writer/editor, as well as a technical specialist. Making high-value technical information available through white papers can strengthen customer relationships. In business, white papers provide information that enables readers to evaluate products, services or technologies. In the IT sector for example, choosing the right technology has become critical to business success. Product evaluation has therefore become more intense, particularly on large-scale projects. Preparing a technical – and financial – justification is an extremely rigorous process. The detailed information available in white papers meets part of that research process, providing documentation that can guide and support decision-making. White papers fall into a number of different categories, including:

- Technology guides – explaining a product's technology, why the technology is important to potential customers, and how it's different from and better than similar technologies.

- Position papers – explaining a trend, or technology.

- Business guides – explaining the business and financial benefits of a product or service.

- Competitive reviews – evaluating a company's products and comparing them with similar competitive offerings.

- Product guides – providing a detailed description and explanation of a product's features and functionality.

- Application guides – describing the application of a product or technology in a particular industry.

A white paper should be a piece of information, not a sales document. A good test is to avoid mentioning your company or your product in the first half of the white paper. If prospects feel that you are trying to sell, rather than inform, the white paper becomes less valuable, and is treated as a marketing message. Your white paper should therefore review problems or needs faced by the reader, rather than describe a product or service offered by your company. You can also maintain the educational value of the paper by talking about your product in generic terms, rather than by brand name – products with these features help to overcome these issues, for example. If you take this approach, you can mention your product name at the end of the document in a 'sign-off' which is separated from the main text of the document.

Although white papers are regarded primarily as a medium for technical readers, it's important to write at the correct technical level. You should not assume that your readers will be familiar with technology, particularly if it is new technology. You should therefore avoid jargon and use charts, diagrams or other forms of visual treatment to clarify complex ideas. Wherever possible, use industry-standard terminology, rather than your company's marketing language. A glossary of terms would be useful for complex products or services. You should remember that white papers may also be read by non-technical readers. If communicating with business decision-makers is one of your important objectives, you must ensure that the content is understandable to a non-specialist audience. Some companies produce several versions of the same paper, with a description on the front cover such as 'this publication is intended for technical readers' or 'this publication is intended for executive decisions makers'.

Because of their important role in purchasing decisions, white papers are now widely available as downloads on suppliers' websites. The Internet has become a key medium for product research and white papers are an integral part of that process. This enables you to distribute information to prospects in a very accessible and cost-effective way. You should make it easy for visitors to find your white papers:

- Create a library listing all the white papers available;
- Place links to white papers on pages where you describe relevant products or industry solutions;
- Place a list of white papers on your press information page.

You can also distribute white papers through companies who specialise in syndicating white papers for a particular industry. These companies offer readers a wide range of material from different suppliers so that readers can compare different offerings.

Managing Knowledge

Generating the content for thought leadership programmes requires careful management of the knowledge and expertise in your company. That means using software to capture the knowledge employees have, storing it in a database and giving employees and customers access to the knowledge.

That knowledge can prove an important differentiator in other ways, such as improving the cost and quality of existing products or services, accelerating new product innovation, or improving productivity by giving employees access to existing best practice, rather than reinventing the wheel. That's why it's important to lead the knowledge management programme from the top. Explain why the company is operating the programme and show how it will help all employees improve their performance. Knowledge management gathers the knowledge that all employees hold and shares best practice across the company.

You can use knowledge management software to capture and categorise the company's existing knowledge. Identify the sources of knowledge such as project reports, proposals, customer records, case studies and other documents. Interview and record information from employees with specialist knowledge or experience. To help employees make effective use of the programme, create a directory of available information and provide access over a secure internal network or intranet. Let employees know about the knowledgebase and explain how to make use of the knowledge. You can use social networking tools to create profiles showing employees' areas of expertise and current knowledge projects. This enables other employees to easily identify colleagues who can help them with specialist knowledge.

Set up a community on your website and encourage employees to submit ideas, tips, best practices and other forms of knowledge. You can also invite suppliers, customers and business partners to submit knowledge on the community page. Collaboration is critical to the effectiveness and productivity of knowledge workers. Set up videoconferencing facilities so that knowledge workers can collaborate with partners in other organisations as well as colleagues inside the company.

To maintain your knowledge advantage, support continuous learning and encourage knowledge workers to attend conferences or other events that will enhance their knowledge. Build relationships with universities and independent research institutes to exchange knowledge and provide additional resources for knowledge workers. Use the knowledgebase to support company growth. Making use of knowledge helps existing employees provide better service to customers or speed up processes such as new product development. The knowledgebase also provides a valuable resource for inducting and training new employees and minimises the risk of lost skills and knowledge when employees leave a company.

Personalisation

The rapid development of data storage and data analysis tools means that it is now possible to know far more about your customers – with information well beyond details of income, spending patterns, service preferences and frequency of use. The information available represents a quantum leap in the ability to profile customers. As a result, the contact between buyer and seller on the Internet is moving towards the ultimate one-to-one experience. Database technology supports a level of personalisation that can deliver highly tailored products and services to specific individuals.

Each time a customer logs on to a website, for example, the database can pull together purchase history and personal preferences as a basis for a highly personalised response. By giving customers a single point of entry, you can increase customer loyalty and learn more about their purchasing patterns. This provides an excellent basis for adding value and for the development of new products.

Building one-to-one relationships involves collecting and using information about actual and prospective customers as a basis for a customised selling approach. This provides an efficient and targeted means of maximising sales. For one-to-one marketing to work effectively, you need the right level of information on customers. You may not always be able to get that level of information on individual customers. However, if you have sufficient information on groups of customers with common needs, you can use techniques such as direct mail to communicate with a degree of precision. As your information on individual customers grows, you can move towards one-to-one communication. One-to-one marketing is designed to:

- Improve the quality of customer service;
- Strengthen customer relationships;
- Maximise the profitability of each customer relationship;
- Increase retention rates for customers;
- Maximise the return on your investment in marketing and customer service.

One-to-one marketing cannot guarantee loyalty, but it can make an important contribution. Customers will only remain if they continue to recognise the value of your products and the quality of your customer service. That means continually enhancing the customer experience.

The more information you have about your target audience, the more precise you can make your campaign. In an ideal world, direct marketing techniques would allow you to communicate one-to-one with every prospect, but in practical terms you are more likely to be communicating with groups who share the same characteristics. This enables you to develop a unique relationship that competitors will find very difficult to match. It can also reduce your marketing and customer management costs by reducing wastage to a minimum.

At the heart of effective one-to-one marketing is a data networking solution that collects, stores, manages and distributes all relevant customer information via a single, integrated customer database. The database is updated from all customer channels and is accessible by all customer-facing staff. The more you know about your customers, the better your chances of increasing their value to your company. Data capture must therefore be an integral part of all your sales, marketing and customer service campaigns. You can build detailed profiles through campaign responses and customer research, and use the latest database and communications technology to manage, analyse and distribute information. Take every opportunity to find out more about your customers so that you can build a real competitive edge, based on one-to-one personal relationships.

Welcome Pages

A key element of the one-to-one relationship is the personalised welcome page on your website. This is a modified home page with personalised messages triggered by identification of a customer and it can boost online revenue as well as reinforcing the relationship. Commentators have identified two types of personalisation 'implicit' and 'explicit'.

Implicit user profiles are created from user behaviour or from user information that can be automatically identified, such as IP address or browser information. The most common form of implicit personalisation is based on the history of a user's clicks or pages they have recently viewed to personalise future content. Explicit user profiles are created when a user registers with a site. The details that the user enters during registration are used to personalise content on a site.

The welcome page uses intelligence in a database to precisely tailor marketing programmes to individual customers. It offers recommendations to customers based on their purchase history and previous searches and reminds customers of their recent searches on site. The page can also include special offers based on the value of customer or their frequency of purchase and may feature fast-track buying mechanism like Amazon's '1-click' with customer details preloaded. For new visitors, it can display messages like 'customers who bought x also bought y and z' to encourage browsing. Personalising a website landing page brings many important benefits:

- More efficient conversion of visitors into customers;
- Increased profit from advertising investments;
- Improved customer satisfaction;
- Higher rate of repeat visits;
- Increased revenue and enhance profitability from higher purchasing levels;
- Deeper insight into customer behaviour.

If you offer personal pages on your website, you can allow customers to add further choices to the profile, using a special checklist. However, it is important to use customer information in appropriate ways. Attempts to increase customer interactions and provide more personalised information have made many consumers concerned about privacy issues. Critics of personalised welcome pages believe that some consumers are suspicious of websites that try to extract information from them. They claim that these consumers avoid website personalisation because of concerns that marketers would misuse the information. It is therefore essential to let customers control the frequency and scope of interaction.

Information should be used to meet individual customer needs. Customers are aware of the value of their information and are only willing to provide it when they see real benefits. That means giving customers control over their data and the way they interact with your company. To build trust, you must allow customers to choose how they want to interact and to use the information that they provide. Once you have customer information, it is important to act on it. Maintain regular contact by sending customers information or special offers tailored to their individual needs.

Website software can be used to automatically personalise the contents on your home page as soon as a visitor arrives. Use phrases that have been proven to work, such as:

- People who bought this item also bought ...
- Welcome back. Based on your history with us, we think you will like ...
- Thanks for your recent order. People who bought what you ordered usually added ...

You can use the information in your profiles to generate recommendations for the products that are most relevant for each customer. As well as increasing direct sales, this approach also results in an increase in the number of clicks per visit so that the visitor stays longer and sees more relevant products. In the longer term, this can improve customer loyalty and frequency of return visits.

Rather than just basing personalised offers on an individual's purchase and browsing history, you can also factor in demographic classifications as well. Offers will begin to change from 'people who bought x also bought y' to 'people like you who bought x also bought y'. This can lead to far more sophisticated and specific marketing activities, with multiple personalised campaigns for each of your products or services. Factors such as the shopper's location, age, sex and occupation can be used to determine the way they are targeted and the offers they receive.

Knowledge Differentiates Service

This type of information also allows you to offer different levels of service to each category of customer. These could include privileged rewards for top customers, incentives for regular customers to spend more or special offers to lapsed customers. Interactive facilities on a website allow customers to design their own customised products. Computers are good examples. The customer chooses a basic model, and then selects features and options from a database. The system provides a price for the customised product, then gives the customer the choice of ordering now or storing the specification on a personal Web page for later modification. The high level of interaction gives customers greater choice and provides you with detailed insight into their needs.

You can also use data on customer preferences as a basis for offering personalised information services. Customers specify the type of information they need, and you alert them by email whenever relevant information is available. Personalisation techniques can also play an important role in building and maintaining trust. Trends indicate that consumer trust in advertisements is declining, with more attention paid to recommendations and advice from friends and other consumers. One study reported that almost 80 per cent of European consumers who regularly use the Internet trust recommendations from family and friends, and almost 40 per cent trust recommendations from anonymous online sources. The same survey reported that, before making their purchases, more than 20 per cent of online purchasers were consulting comparison sites or community pages where other consumers write about their experiences with products and the companies that supply them.

You can build trust and encourage stronger commitment from your customers by enabling them to communicate their opinions, write reviews and create their own personal lists on your site. You can also inform your customers about other customers who have the same needs and preferences, describing the products they bought, the reviews that they wrote and other observations.

A growing number of websites feature simplified purchasing systems that allow pre-registered customers to bypass a conventional shopping cart and online checkout by purchasing a product with a single mouse click. The Web server recognises the purchaser and relates their order to specific credit and shipping information previously submitted by the purchaser and stored on the server. This adds even further value to the personalised welcome page as well as reducing the risk of abandoned purchases.

Using intelligence to shape individual customer responses creates even greater revenue opportunities with high-value visitors. An example is a pay-as-you-drive insurance service that employs existing technology such as satellite navigation linked to smartcards to create an innovative model that tracks customers' cars and charges them according to where they drive. This breaks the mould of driver's insurance and enables insurers to reach marginalised markets, like people who don't drive very far in a year. This insight adds value, improving service to customers even further.

Summary

Knowledge helps you build stronger relationships with customers by empowering them with information, building trust and providing you with insight that enables you to personalise products, communications and the level of service you offer.

8

Manage the Market

To manage your position in the market, it's essential to understand and use your competitive advantage. The factors that differentiate your company could be based on product performance, price, quality, capacity, delivery time or financial stability. The starting point is to identify your competitive advantage and build those advantages into customer communications and marketing programmes.

Competitive Advantage

Every company has a set of resources and capabilities that it can use to build a competitive advantage. The stronger those resources and capabilities compared to the competition, the bigger the advantage. However, where there is no real differentiation, competitors can easily replicate your products and services, creating a commodity market where none of the players has any specific advantage. These are examples of resources that competitors would find it difficult to replicate easily:

- Patents and trademarks;
- Proprietary knowledge;
- Customer base;
- Corporate reputation;
- Brand equity.

However, as well as having resources, you must also have the capability to use them effectively and profitably. This is derived from business processes, quality procedures, organisational structures, corporate culture, recruitment and training processes, and relationships with suppliers and business partners. Together, they give your company the capability to operate in a way that delivers success through innovation, efficiency, quality and customer focus. These capabilities are intangible and can be even more difficult for competitors to replicate.

You can use your resources and capabilities to either lower your cost structure or differentiate yourself through superior products. You also need to decide whether to use your strengths across the whole of your chosen market or to focus on a specific sector where the combination of resources and capabilities gives you greatest advantage. If you don't have the scale or resources to compete across the board, you can be successful by concentrating on niche markets. You can also improve competitive advantage by drawing on the resources of partners in your supply chain.

Sustainable competitive advantage is vital to long-term survival and profitability. To achieve this, your company must be responsive to changing market and competitive conditions and innovative so that it can quickly develop new products and services to meet those changes. Competition on a global scale and the emergence of new forms of competition mean that you must be adaptable and able to change quickly to maintain competitive advantage.

You can adopt four generic business strategies to gain competitive advantage:

- Differentiation in a mainstream market;
- Cost leadership in a mainstream market;
- Differentiation focus on a niche market;
- Cost focus on a niche market.

Differentiation – This strategy involves focusing on specific criteria used by buyers in a market and positioning your company as uniquely qualified to meet those criteria. Typically, this type of product would carry a premium price that reflects the higher production costs, tighter quality processes and extra features provided for the customer. Communication is based on building customer preference for a superior brand and justifying the premium price.

This type of strategy is suitable for products like high-specification consumer electronics or gold standard services.

Cost leadership – Companies following this strategy aim to become the lowest-cost producer across all sectors of their chosen market. This strategy is common to large corporations who offer commodity products with little differentiation that appeal to large numbers of customers. Where price is perceived as an important factor by customers, a low-cost leader with a significant cost advantage over competitors is likely to achieve a large market share. This type of strategy is suitable for mass-market products.

Niche differentiation – With this strategy, you aim to differentiate within small niche market segments where you can identify specific customer needs that are not met by the mass-market offerings of competitors. To succeed, it is essential to ensure that customers really do have different needs and wants, and that there are sufficient customers to support a profitable sales and marketing operation. This type of strategy is suitable if you offer specialist services, for example.

Niche cost leadership – With this strategy, you concentrate on building a cost advantage in a small number of niche segments. You could offer basic or lower specification versions of mass-market products and services that appeal to price-conscious buyers. This type of strategy is suitable for retailers and manufacturers who concentrate on discount or own-label products.

Differentiated Offers

With a differentiated marketing strategy, you develop different offers for individual market sectors, rather than a single offer for all sectors. That strategy enables you to meet the specific needs of customers in each sector, increasing the chances of marketing success. You can differentiate the offer in a number of ways: by modifying the product, by offering different level of service or by offering products through different channels. Although a differentiated strategy may increase sales, it may also require higher marketing costs to reach each sector effectively.

Differentiated marketing meets specific customer needs, rather than attempting to cover all bases. Offering the same software product in two versions – for professional users and for home users – enables the supplier to meet the price and performance needs of each sector with only minor modifications to the core product. That strategy enables you to compete effectively in niche sectors where competitors do not have a suitable offering. If you operate in a number of niche sectors and build strong market share in each, you can increase overall sales compared to the sales you would achieve by trying to compete with undifferentiated offers across all sectors. Differentiated products also appeal to retailers and distributors, helping you to attract and build a more effective distribution channel. Retailers benefit from dealing with a single supplier, but have access to a range of products that appeal to different groups of consumers.

A differentiated marketing strategy enables you to respond quickly to changes in the marketplace without major investment. For example, changing the distribution channel from retail outlets to the Internet enables you to meet the needs of customers who prefer the convenience of online purchasing. Offering a consumer technology product with a free home installation service enables you to meet the needs of customers who do not feel confident about installing the product themselves.

Customer Base

Controlling your customer base is an essential step in managing your position in the market. Having an installed base gives you even greater control. A customer base is a set of records that list all of your customers. An installed customer base is a record of the numbers of a specific digital product installed on customers' corporate information systems. A customer base provides general information for planning future sales and marketing programmes. An installed customer base provides a specific platform for building future sales of upgrades to the installed product or related products and services. An installed customer base also acts as a barrier to entry for competitors.

A customer base lists all of your customers, large and small, including those that have bought any product from you at any time. Some customers may purchase regularly, others occasionally. You can use information on the customer base to plan future marketing programmes. Information on individual transactions, purchasing frequency and purchasing patterns enables teams to predict the long-term value of customers as a basis for allocating your marketing resources. Annual revenue derives from sales to the complete customer base. You can grow revenue by targeting new prospects to increase the customer base, or by selling more to the existing base. To maintain control, you must also protect your existing customer base as the cost of dealing with existing customers is lower than the cost of acquiring new ones.

An installed customer base gives you a strong competitive advantage, compared with companies whose customers buy products that competitors can match. Customers that have installed a digital product, integrated it with existing systems and trained their employees to use the product are likely to be reluctant to change to another supplier. The cost of switching suppliers is high and this leads to a situation of vendor lock-in. If you have a large installed customer base, you can also benefit from network effect. This effect occurs when a large number of companies in a market are unwilling to switch to an alternative product that is incompatible with the product in the installed customer base. The network effect also acts as a barrier to entry to competitors.

You can use the installed customer base to offer customers product upgrades and other related products and services. The compatibility of the new products simplifies changeover and reduces any associated costs. Companies marketing a range of products to their customer base do not have the same benefit of compatibility. Customers can choose from a range of competitive alternatives without incurring switching costs.

Enterprise Markets

The size of customer you target can also affect your ability to manage your position in the market. Technology companies use the terms *enterprise* and *small and medium business* (SMB) to identify two distinct market sectors for their products or services. The enterprise market is the sector populated by large organisations. The threshold between SMB and enterprise markets is based on number of employees or annual turnover. Some marketing teams make further distinctions between large and small enterprise sectors.

The terms enterprise and SMB are also used in market segmentation. Some technology companies use vertical or industry sectors to identify target markets where they can differentiate themselves through specialist market knowledge or product characteristics. Others combine the categories, for example marketing products to large enterprises in the financial services sector or marketing services to small healthcare enterprises.

Enterprise markets offer your company the opportunity to sell higher volumes and earn more revenue than you would by marketing the same products to SMBs. However, enterprise customers may be able to negotiate large discounts based on their purchasing volume, a factor that would reduce profit margins on enterprise sales. Purchasing decisions are complex in enterprise markets. In SMBs, a purchasing manager makes buying decisions, with input from specialists such as technical managers for more complicated product decisions. Enterprise purchasing decisions involve a number of different decision-makers, as well as people who influence the decisions. A decision-making team for a major purchase, such as a new manufacturing control system, might include the head of purchasing, chief executive, finance director, manufacturing director and information technology director.

Companies that have high-profile customers in the enterprise sector sometimes use the term enterprise class to describe their products and services. The implication is that enterprise customers have demanding standards of quality or reliability that the supplier is able to meet. On that basis, other customers can feel confident that the product will meet or exceed their requirements.

Your company faces higher levels of risk in enterprise markets. If revenue from a small number of enterprise customers represents a high proportion of your turnover, the loss of just one major customer could have a damaging effect on the company. To protect your enterprise customer base, you should utilise a system of key account management. A team of account managers, customer service staff and technical staff focus on building strong relationships with the customer and creating a partnership that competitors would find hard to overcome.

Market Agility

Speed, flexibility and agility are essential requirements for managing the market and responding to change. Agile companies have four characteristics – visibility, speed, flexibility and scalability. Visibility enables you to anticipate change. Speed is essential to develop and deliver innovative products and services and strengthen your market position. Flexibility means you can make rapid adjustments in response to changes in market demand. Scalability makes it possible to align your people, resources and capacity with a changing environment.

Insight into the marketplace enables you to modify your strategy in response to change. Set up a process to monitor changes in the market and identify trends that might impact your company. Analysts' reports and forecasts provide a useful starting point. Combining research information with company data helps you build a more specific picture. Ask the sales and customer service teams to report any significant changes in customer requirements. Arrange review meetings with major customers to discuss their changing needs and challenges. You should also build a network of trusted partners. If you cannot meet changing market demand from your own resources, you can work with partners that offer suitable products, technology or manufacturing capacity.

Utilise cloud computing to scale information technology resources. Market change may require a significant increase in computing resources to deal with increased demand. To avoid the time lag associated with internal investment, an organisation can utilise cloud computing – renting additional capacity on massive servers based in data centres run by external service providers. This provides immediate access to essential resources and reduces time to market for new business opportunities.

The pace and level of competition may become so fierce that it is necessary to transform the business. As an example, the development of information technologies has given smaller companies the ability to compete effectively with much larger organisations. In some markets, such as telecommunications, deregulation has broken up the monopolies of large organisations and given their smaller competitors new business opportunities. In this scenario, both large and small competitors may need to transform their business model to adapt and survive in the new marketplace. That means identifying new opportunities and strategies to build or protect competitive advantage, focusing on new product development, growing market share, strengthening customer loyalty and eliminating communication barriers.

Telecommunication companies market their products and services to consumers, business customers and other service providers. The aim of telecommunication marketing is to generate the highest level of revenue from a network and strengthen customer loyalty. In consumer markets, companies focus on selling bundles of products to customers, while in business sectors, they aim to build strong relationships as a channel for marketing additional services.

In consumer markets, telecommunication companies can utilise the broadband access that many customers have taken up to offer what is known as the triple play – telephone, video and Internet. This enables them to increase revenue and customer loyalty, as well as protecting their customer base against competitive activity from low-cost telephone service providers and indirect competitors such as cable network companies.

An increasing number are marketing their services to business customers using consultative selling techniques. They work with customers to discuss their communication needs and provide advice on the services that meet those needs. The rapid changes in communication technology make it important to educate customers on the relevance and benefits of new technologies. Discussion papers, seminars and other events form an important part of the marketing programme.

The wholesale market is also important, with network owners offering service providers, such as smaller telecommunication companies, online video publishers and content providers, access to the network to carry their traffic. These are called carrier services; they increase revenue for the business and enable other providers to extend their networks without building additional infrastructure. To market carrier services, telecommunications companies communicate the benefits of a network with the reach, performance and capacity to meet customers' needs.

To increase revenue and return on investment in the network, they are developing new products and services. The latest multi-service networks, based on Internet Protocol (IP) technology, enable operators to offer a much wider range of services on the same network than traditional dedicated voice networks. They are using the capability of IP networks to build video, data, Internet and voice services over the same infrastructure. The increasing importance of mobile telecommunication threatens the voice revenue that was the mainstay of traditional telecommunication providers. Many are responding by developing their own mobile offering or working in partnership with other mobile operators to offer joint services.

Competitive Analysis

To maintain control and counter competitive actions, you need to identify the advantages you and your competitors have over each other. By monitoring and analysing competitive activity, you can predict competitor behaviour – what they are planning and how they might react to moves that you make. Controlling competitive activity can help to meet many different aspects of your marketing strategy. These are some examples:

- Maintain sales to existing customers;
- Protect revenue streams;
- Win business from new accounts;
- Counter competitive activity.

Competitor information helps you to identify how you can protect your most important business and, more positively, how you can strengthen your position with customers in situations where your competitors are currently holding a larger share of the business than you. Competitor intelligence is freely available from the Internet, the press and other published sources. Information from these sources can provide a valuable starting point for developing more detailed competitor profiles. The sales force can obtain competitor information from many different sources. By talking to customers, they can find out about competitors' direct sales calls, direct marketing campaigns, special offers and new developments. They can also obtain similar information from retailers or distributors. The sales force may be able to provide a profile of your most important competitors.

Information is readily available from publications, the press and the Internet that can help you gather intelligence on different aspects of competitors' business, including:

- Main markets;
- Customers;
- Resources and financial performance;
- Product range;
- New products;

- Plans for growth.

Corporate brochures and annual reports are the best sources of published information. You can obtain copies from exhibitions, customers or as downloads from competitors' websites. Maintain a file of press cuttings on your competitors' activities using their trade publications as a source. You can use a press cutting agency to gather material for you. Ask them to provide you with any cuttings that include the names of your competitors. Published industry surveys can provide a useful insight into purchasing patterns in different marketing sectors. You can obtain information on competitors, market share and industry trends. Industry surveys are often sponsored by magazines, industry associations or groups of manufacturers.

Competitor websites can provide valuable information on their resources, plans and capabilities. As well as checking the company information and product pages, you should read any customer case studies on the site and monitor the news section to find out about new developments. The Internet is also a valuable source of other competitor research material. Industry newsletters or summaries of published research could prove a valuable starting point. Trade publications are increasingly available in electronic editions, making it easier for you to monitor the press. Competitors' exhibition stands can be valuable sources of information. Most companies only participate in exhibitions that are important to their current or future business plans. Frequently, they choose to exhibit because they have a new product to announce.

Analysing competitor marketing strategies will help you to respond to their activities. By monitoring competitor advertising, promotions, exhibition presence, press activities and Internet information you can assess possible strategies. These are some of the possible scenarios:

- Heavy advertising expenditure could indicate a competitor trying to win greater share.

- Price promotions may indicate that your competitors want to be perceived as value-for-money suppliers.

- Press announcements about new production facilities could indicate that your competitors are trying to significantly increase their business.

- Announcements about new branch or dealership openings could mean that competitors are expanding into new territories.

Marketing publications can be useful sources of information on competitor marketing activity. They publish news of recent campaigns, league tables of expenditure by different companies and analysis of company strategies. The recruitment and appointments sections of marketing publications can also highlight changes in strategy.

If you do not have the internal resources to monitor competitive activity, you can use an independent research company to carry out all the tasks outlined above. You can also ask them to carry out research into customer attitudes to competitors. Customers may be more willing to provide this information to an independent organisation. You should be cautious in acting on competitor intelligence. Published sources can only provide a partial picture; more strategic information is likely to be confidential. This means that you may make incorrect assumptions in planning your response to competitor action.

Barriers to Entry

One way to control competitive activity is to develop strategies that create barriers to entry in markets where they compete. The strategies include cost, technology, investment, knowledge, customer relationships and switching costs. If entry barriers are high, it is difficult for new competitors to enter a market without major investment and significant risk. Markets with low entry barriers allow competitors to enter easily, resulting in greater choice for customers.

Low-cost production is the most important barrier. Organisations in countries with low labour costs, particularly those in emerging economies, can offer prices that high-cost producers find difficult to match. In technology markets, large-scale capital investment is essential to develop new products or manufacture in high volume to meet market demands. In the microprocessor industry, for example, new entrants would have to make massive investments to match the clean room environments of existing suppliers.

The technology sector features an entry barrier known as the network effect. Companies spend heavily to build a large base of customers who use compatible products such as word processing software to interact with each other. That customer base creates a network that manufacturers of incompatible products would find difficult to enter. If you hold product patents, you can prevent market entrants competing directly with similar products. Patents can have positive and negative effects on a market. They reward and encourage innovation that requires high levels of investment, for example in the development of new biotechnology solutions.

Where raw materials or other resources are scarce, companies aim to secure supplies to protect their business and create barriers to market entry. China, for example, supplies the majority of the world's stock of rare earth metals used in the manufacture of high technology products such as mobile phone handsets.

You can use switching costs to protect your most important customers and create barriers to market entry. Switching costs take a number of forms. Customers might take out an annual contract for a service and find a lower price part way through the contract. Switching suppliers may incur a penalty charge. When you train customers' staff to use your products, customers would face further training costs if they switched to a competitive product.

Summary

The more control you have over the market, the stronger your competitive position. You should identify the factors that differentiate your company and use them to respond to market opportunities and competitive threats. An installed base strengthens your market position, but you must be able to respond quickly to changing market conditions. Understanding the competitive challenge is key to overcoming threats, while barriers to entry can help to protect your market position.

9

Develop Effective Channels to Market

Distributors play a key role in technology markets, providing your company with sales, marketing and customer service functions support. Distributors can take the form of franchised or wholly owned organisations working for one manufacturer, or they can operate as independent agents, acting for multiple suppliers. They enable you to reach geographically remote customers or niche market sectors that you could not support cost effectively using your own internal resources. Improving distributor performance helps you sell more products or enhance levels of service to your smaller customers.

Appointing distributors allows you to develop a flexible marketing strategy, using the most appropriate channel, or route to market, to reach different market sectors. As well as using a direct sales force and intermediaries, you can also exploit online channels, marketing products via the Internet or introducing digital distribution of suitable products and services.

If you sell products or services to a number of different market segments, you may not have the resources to develop your business in all segments. By working with distribution partners that have specialist knowledge of the segment, you can operate the equivalent of a niche marketing strategy. You can support the distributor and improve penetration in the segment by providing market development funds or running joint marketing campaigns.

Distributors generally deal with a manufacturer's smaller customers, leaving your sales force to focus on key accounts. They can also provide an after-sales service to all of your customers, large and small, by holding stocks of spare parts and providing a local technical and maintenance service.

Working in partnership with distributors opens wider geographical markets. Setting up a dedicated global distribution network, for example, would require a massive investment of time and resources. Using established networks of distributors that offer global coverage provides a rapid, low-cost solution. Distributors make it easier to enter new markets. You can draw on the distributor's market penetration and knowledge, rather than committing sales and marketing resources to build a presence from scratch. They can also offer coverage and specialist knowledge of small niche markets that provide an additional outlet for low-volume products in your range.

Distributor Management

The structure of your distributor channel is important to marketing success. Some companies operate different levels of distributor, with programmes that enable distributors to move up the structure and gain additional support benefits. A tiered programme can help you attract high-quality distributors and improve performance so that distributors deliver a high standard of service to your customers.

Selecting a distribution channel is an important aspect of building a competitive advantage for your business. Distribution channels include your own direct sales force, retailers, distributors and the Internet. The right distribution channel ensures that customers in different locations around the country, or around the world, can buy your products and get the right level of service from your company. To select the right distribution channel for your business, you need to consider a number of factors, including location and reach, skills and resources, management costs and degree of control. The channel you choose must be easily accessible for customers and prospects, wherever they are located. If you sell products locally, choose a retailer or distributor in your area who knows the local market. If you are expanding and want to sell your products in other states, choose a distribution network that provides coverage of your chosen markets.

You want your customers to receive the best possible service. That's not easy when you are dealing with customers through a third-party distribution channel. If you sell high-value, complex products to business customers, you need sales people with good product knowledge and the ability to build relationships with customers. A direct sales force may be the right solution if your customer base is small. If you have a large group of business customers around the country, look for a distributor network selling similar products. They will have the skills and knowledge to sell your products, particularly if you provide their sales team with product training.

Using a distribution channel doesn't just give you a method of delivering products to your own customers; it could also give you access to a much larger customer base. Look for a distributor or retail network that sells products that are complementary to your own. By offering sales incentives to the distributor, you can tap into their customer base and grow your own business.

Establish a tiered structure for your distributor network that provides partners with increasing levels of support as they achieve higher ranking. The Cisco Channel Partner Program is a good example, offering Registered, Authorised or Certified Partners different levels of benefits such as certification for technical staff, access to market-leading products and services, marketing and financial support, and opportunities for business growth.

Delivering a certification programme for distributor technical staff develops essential skills in deploying and supporting your products. The biggest problems for technology vendors occur before, during and after the deployment of each technology. The programme can include distance learning programmes as well as practical, hands on training workshops on your site or in educational institutes. By working with distributors in this way, you can help them develop their business so that they can move up the structure and deliver increasingly higher standards of service.

Availability of up-to-date business, technical and marketing information is essential to the operation and development of a distributor network. Effective communication throughout the distributor network helps improve the efficiency of your marketing operations. Communication enables all members of the distribution network to share essential market and operational information so that they can respond quickly and effectively to changes in the marketplace. You should provide essential market information that helps distributors plan their own marketing and investment programmes. If you have acquired new customers, launched new products or entered new market sectors, the changes will have an impact throughout the distributor network.

Regular marketing communications also build teamwork by keeping all parties informed on developments that affect their operations and by helping them to grow their own business. By bringing together all parties through communications, you can build an efficient, responsive network that operates as a single, coordinated unit. Providing a single integrated source of information creates a distributor network that can respond rapidly to change or new opportunities. The problem is that as the distributor network grows, the task of trying to communicate with members who might be located anywhere in the world over a variety of different communication networks, devices and standards gets more and more complex. The communication network should therefore be based on open standards so that all members can communicate regardless of the nature of their internal information technology systems.

Set up information portals for distributors to strengthen relationships and improve productivity. The portal should include a document resource library, marketing collateral, e-learning tools, certification management tools and other important business development material. Incorporate a secure forum to allow distributors to share information and build a sense of community. Putting the material in a portal reduces distribution costs and ensures that material is available to the entire network. However, the onus is on individual distributors to access and use material, so it is important to monitor portal activity to measure how content is being used. Sharing market knowledge is a two-way process in distribution channel marketing. The distributor brings front-line knowledge of local customers and market requirements. You provide product knowledge and information on sales and marketing techniques to support the distributor's team.

Control is an important factor is distributor management. It is the people in the channel who control the relationship with your customers, so the right choice is critical. Is the distribution network selling products that are competitive to yours? Can you motivate the network to give preference to your products? Is the network willing to operate to the standards of customer service you set? Do you have the resources to monitor and manage the performance of the distribution channel?

Marketing through distribution partners can be a double-edged sword. Companies that rely on distributors to sell, support and market products can find it difficult to control channel performance. Distributors can change focus or switch allegiance to other manufacturers. That can have an impact on revenue from the channel and the quality of service to end customers. Channel support is no longer a simple business practice, it requires tight management and focused, directed support.

While you benefit from the reach that distributors give you, you can lose control of the message that end customers receive. You can help to control the consistency of marketing messages by providing clear guidelines on communications strategy, as well as branded marketing collateral. Setting up an approval process for distributor communications also helps to ensure consistency.

You can achieve control by imposing quality standards, providing training and developing standard procedures for distributors' customer-facing processes. You should also aim to control distributors' sales and marketing activities so that you can maintain a greater share of their business. Co-marketing programmes, tiered sales incentives and strict targets keep distributors focused on developing business for you, rather than competitors or other companies that they represent.

Distributor Marketing Support

Marketing support is crucial to improving distributor performance. You should set up a team to work with local distributors to help them develop their strategy and plan their programme of sales and marketing activities. The team would arrange regional meetings to help local distributors identify the right strategic marketing mix and messaging. To help them implement the programme, the team can recommend local approved marketing agencies that work within the framework of your overall marketing strategy and can ensure that local marketeers are up to date with the latest campaigns and creative strategies.

Local markets share many common characteristics, as well as clear local differences. Sharing experience and best practice can help all local markets, so consider setting up a local market forum. The more mature markets may have case studies and proven campaigns to share, while younger territories can share their experiences. This experience can help to identify the support requirements for other territories or groups of distributors.

You can design programmes to help distributors grow their own business and make it easier for all parties to work together. A business education programme will help distributors keep up with the latest developments and maintain their product and business skills. Courses could cover management development, product marketing skills, industry knowledge and sales and marketing tools.

Support can take the form of training, provision of marketing materials or join marketing initiatives. Distribution channel partners offer manufacturers local marketing knowledge, access to an established customer base and a sales force that can add value to the marketing of a product. To help your channel partners deliver the added value, you must provide high levels of support in the form of product information, training, marketing collateral and funding for campaigns. You may also appoint channel marketing managers to work with channel partners to develop business.

Provide authorised distributors with branding material to use in their own customer marketing and communication programmes. Emphasise the benefits to their business of using a strong, widely recognised brand by explaining that the award of a brand demonstrates to customers that the distributor has the capability to meet your high standards. Back up the branding programme with customer communications recommending the use of authorised distributors.

Produce product information and sales guides for distributors' sales and marketing teams. Produce guides or sales bulletins on specific market opportunities. Create a website that allows vendors to access and download product information and other marketing resources. Offer selected distributors the opportunity to participate in co-marketing programmes, concentrating on distributors with the potential to deliver good results from the programme. Offer financial support for an activity such as a direct marketing campaign or provide marketing materials such as advertisement templates or product brochures.

Work with vendors to help them develop their business. Help vendors to develop new services for their customers by providing guides and technical support. If you operate a tiered structure for your network, you can provide distributors with increasing levels of support as they achieve higher ranking.

By offering local outlets a choice of programmes, you can tailor your marketing to suit local market conditions. The programmes might include:

- Local advertising;
- Direct mail;
- Product literature;

- Display material;
- Merchandising material;
- Special offers;
- Public relations support.

You can produce a guide to support programmes which enables the local outlet to select programmes that allow them to develop their own promotional strategies. The guides should:

- Explain the scope and benefits of individual programmes;
- Describe the support material available to operate the programme;
- Explain how to order support material;
- Provide guidelines on running the programmes.

Because there are so many possible variables in different markets, you should develop localisation kits that will enable you to customise your creative work to incorporate the most relevant messaging, images, offers and creative treatment while retaining the value and strength of your brand. This approach gives you real creative and commercial flexibility. The key stages in localisation are analysing and refining messages to reflect levels of market maturity and awareness or local issues, and selecting appropriate images or graphics to suit the local culture.

The essence of an effective local support programme is that local distributors understand their customers' needs and communicate with messages tailored to that market. The most powerful medium to achieve that is direct marketing – a powerful weapon in the hands of local outlets because it enables them to talk to their customers on their own terms. The key to the success of local direct marketing programmes is detailed knowledge of the local customer base so that the offers and information can be tailored to local marketing programmes. The most efficient way to handle this is to maintain a central database of all local customers and use database management techniques to manage the mailing list. Local distributors are unlikely to have the sophisticated equipment needed to carry out database management operations and the exercise can be carried out more efficiently on a central database. Information for the database can be gathered from a number of sources, including:

- Local customer sales records;
- Replies to advertisements;
- Responses to special offers or invitations;
- Market research.

Local or regional advertising campaigns can be customised to suit the needs of the local market. Support can be delivered in a number of forms:

- Funds to enable local outlets to produce and run their own advertisements.
- Contributions to the cost of joint supplier/local outlet advertisements.
- Contributions to the cost of advertisements run by regional groups of outlets.
- Production of national support advertisements which incorporate local information and which are run on a regional basis.

The level of financial support depends on the funds available for local support and your local distributors' own budgets. For example, many independent outlets have substantial advertising budgets of their own and utilise the suppliers' budgets to supplement their own or to run specific campaigns. Other smaller outlets or franchised outlets without their own budgets rely entirely on the supplier's contribution to run local campaigns. Because of this, the question of financial support is usually subject to negotiation.

You can also supply support in the form of complete advertisements, logos, artwork or photographs for inclusion in the outlet's own local campaigns. As you are more likely to be concerned about the consistency of local advertisements than the local outlet, you should issue clear guidelines on the use of different elements of corporate identity. Many suppliers provide advertising standard manuals which give examples of layouts for different sizes of advertisements, explain the position and size of the company name and logo, list the typefaces to use and include sample advertisements for guidance.

Alternatively, you can offer local outlets a central advertising service. This support policy enables you to offer local outlets consistent professional advertisement standards, with local information such as name and address, map, priced offers, product variations and special offers incorporated. The local outlet benefits from national advertising and strong branding, but they have advertisements that suit the local market.

International Distribution

Distribution plays a key role in the success of international marketing strategy. It ensures that your company can deliver products and services to international customers efficiently and cost effectively. Members of the distribution channel also provide an important local marketing resource that enables you to increase market share or enter new markets with the backing of local knowledge and established business relationships. If you are developing a global branding strategy, you can use a distribution channel to extend your brand into new territories.

You can manage international distribution in a variety of ways from operating wholly owned subsidiaries to partnerships with independent or franchised organisations. Using independent distribution organisations eliminates the direct costs of building and operating a local infrastructure. However, the indirect approach gives you less control over the day-to-day marketing and customer service operations in the local market.

Companies operating globally generally manage their markets by three main areas – Americas, EMEA (Europe, Middle East and Africa) and Asia. The distribution strategy can vary by territory within those broad groupings. If you offer different products to different territories, you will need to select distribution channels with the appropriate level of sales, technical and support skills in each territory. Efficient information exchange between you and your distributors is essential so that all parties can respond rapidly to change or new opportunities. Internet communications and the availability of secure global data networks make it possible to link partners located anywhere in the world.

Your international distributors can operate as basic logistics partners or offer a wider logistics and marketing service, using their local knowledge to help companies increase market share. You can improve the effectiveness of local marketing by providing product training, campaign funding and communications material adapted for use in individual territories. If you have strong established domestic brands, you can use the distribution channel to develop a global brand strategy. By encouraging distributors to adopt some or all of your brand names and identity, you can build on your reputation in wider markets. This can benefit both parties, driving business to the distributor and reducing brand-building costs in the local market.

Multi-Channel Strategy

A multi-channel strategy offers your customers a choice of ways to buy products. A true multi-channel strategy covers purchases from a distributor or retailer, online purchase from a website, telephone ordering, mail order, interactive television, catalogue ordering and comparison shopping sites. The aim of a multi-channel retailing strategy is to maximise revenue and loyalty by offering your customers choice and convenience.

A successful multi-channel strategy offers your customers a consistent quality of experience, whichever channel they use. The customer's experience in buying a product influences their perception of your brand. You must ensure that all customer-facing staff in your contact centres, retail outlets, order processing departments and website development understand and comply with your company's customer service standards. Policies such as returns and delivery charges should also be consistent across each channel.

You can use the data available from multi-channel strategies to identify customer preferences and maximise lifetime revenue, according to IBM. The risk in a multi-channel strategy is that customers' information and profiles can become fragmented as they use a number of channels. In an integrated strategy, purchasing data is consolidated into a single database so that you get a 360-degree view of your customer.

IBM reported that an informative website is a key element in a multi-channel retailing strategy. A high proportion of shoppers conduct their pre-purchase research on the Internet, although they may use other channels to make the final purchase. The definition of multi-channel retailing must therefore take account of each stage of the purchasing process.

When you operate a multi-channel strategy, you have the option of charging the same price for a product across all channels, or offering your customers different prices, depending on their channel choice. Offering customers website prices that are lower than store prices is common practice and reflects the lower overheads in website operations. Your customers also have a choice between lower online prices and personal service in the store.

A multi-channel strategy gives you the ability to expand your business on a national or global scale without investing in further physical stores. An e-commerce website can be designed to operate in multiple currencies and with local content to attract customers in many countries. You can also conduct business across multiple time zones with no restrictions on opening or closing times.

Summary

Your channel strategy makes an important contribution to marketing success. The right channel gives you access to a wide customer base and enables you to deliver a level of service that you would not be able to provide from your own resources. To use a channel effectively, you have to manage it carefully and provide the right level of support. Channel support helps your distribution partners grow business for you and deliver the right level of service to your customers.

10

Maintain a Culture of Innovation

Innovation is critical for companies who want to remain competitive in the long term. It's a key discipline within product development and it underpins quality customer service. Although many companies recognise its importance, it can be difficult to identify the sources of innovation and create an innovative culture. Although innovation is likely to originate within a product development group, the most productive solution is to build a culture where people throughout your company are proactive in developing innovative ideas.

It's important to distinguish between innovation and invention. Invention is the process of discovering things that have never been discovered before. In business, innovation is the discovery of new ways of creating value. To achieve that, employees don't have to be working in product development or a research team. Their day-to-day experiences of working with customers or overcoming challenges can highlight a better way of dealing with an issue. So although few people are likely to be inventors, anyone can be innovative.

Sourcing Innovation Ideas

An important element in building company-wide innovation is to encourage small improvements, rather than looking for the 'big idea'. For example, you could ask employees to look at different aspects of the business. How can the company improve delivery to customers, save time on product development or promote products in a different way? The ideas can be very simple and probably only a few will be useful. But, this approach encourages people to contribute. The challenge is to build a culture that encourages all employees to contribute to innovation. This can be difficult when employees and managers are focused on day-to-day pressures or when departments are used to working in isolation.

A security company recognised this problem and decided to move forward by progressing in stages. First, it put together a team of people who could drive innovation. This team was responsible for raising awareness of innovation, building any necessary infrastructure, creating the training materials and plans, and developing a process for managing innovation. The team included people from all departments, particularly customer-facing departments, and marketing was at its core. The aim was to ensure that all departments recognised their responsibilities and opportunities. The company recognised that a specialist team was an essential first stage, but it should be seen only as a starting point. After a period of time, innovation should be the responsibility of everyone in the company, not just a specialist group.

They soon found that members of the innovation team were taking on different roles and responsibilities, depending on their experience and attitudes. People regarded as true innovators tended to do things differently, breaking the rules and ignoring traditional ways of doing things. More conservative team members were prepared to adapt and make improvements, but worked within the rules and accepted the status quo. The group also included people who were good at detailed planning and others who lacked planning skills but were good at implementing solutions and making things work.

If you're planning to spread the word on innovation, try to get a balance of those characteristics. An effective team can achieve great results, but one of their most important roles is to spread the load and build a commitment to innovation throughout a company. For that security company, the next stage was to create an 'innovation community' – a larger group that represented the specific needs of departments, as well as championing the needs of customers, business partners and other stakeholders. The community operated informally, but built relations with the innovation team, identifying needs and communicating those needs to the team. Community members carried out a range of important tasks to support the company's innovation drive. One group delivered workshops to explain and encourage innovation within different departments. Another stimulated innovative ideas through describing best practice and running brainstorming sessions. And, to capture the benefit of those sessions, they prepared case studies which proved invaluable for later projects.

Support Innovation

If you're aiming to create this kind of environment, it's important to provide facilities to support the innovation community. You can set up a 'virtual community' on the Internet or your company intranet to support communications and collaborative working between people with a common interest. Facilities to support the community could include newsletters, discussion groups and information. An online discussion group gives users the facilities for posting messages on the innovation community site. The messages should represent helpful information and may include requests for help or further information. Some sites set up facilities for feedback or review, introducing an opportunity for objective, independent comment.

The aim is to encourage other members of the community to suggest answers, provide help or contribute to the discussion of a specific issue. Discussion groups help to strengthen the relationships that are essential to a company-wide culture of innovation. If the community is successful, they can spread understanding and commitment to innovation throughout the company. The objective is to make innovation an everyday activity. By this stage the responsibility for change and innovation no longer rests with an individual. Employees take over responsibility themselves and see innovation as an essential part of their jobs. You can encourage good contributions by recognising and highlighting the ideas that get accepted. Put the best ideas onto the innovation website and involve other employees. The key is a commitment from top management to implement the best ideas. If your employees see the prospect of their ideas being accepted and rewarded, they will contribute.

It's clear from these examples that marketing works best when it involves everyone in a company. Your natural partners are the sales force, but working in the background are all those people whose jobs bring them into direct or indirect contact with customers. Encourage them, involve them and keep them informed on what the company is doing and where it's going. When people make a positive contribution to success, recognise their achievement and reward it. It may not be marketing in the conventional sense, but it will certainly add a valuable resource and edge to your efforts.

Collaboration Tools

Growing companies make the most productive use of the skills and knowledge of their own people and their business partners. Using collaboration tools like email, teleconferencing, videoconferencing and instant messaging, you can create 'virtual teams' that bring together the right people at the right time, even when they are in different locations. Collaboration drives growth by enabling teams to tackle business challenges more effectively, speeding up decision-making and improving key activities like new product development. The tools and technologies to support collaboration across different locations include instant messaging to speed up team communications and email to exchange drawings, documents and project information. Teams hold meetings by teleconferencing, videoconferencing or Internet to reduce travel time and cost. They can use websites to create a single source of project documentation coupled with web tools to enable team members to view and comment on project documents. Vital information is no longer hidden in departmental 'silos'.

Effective teamwork is key to growth. Team members get the information they need from colleagues and draw on each other's skills to push growth projects forward. With collaboration tools in place, you can easily create focused teams from different departments like design, engineering, manufacturing, marketing, sales or finance. These tools help to create a 'virtual project room' where team members can focus on their growth targets by holding planning and progress meetings, brainstorming sessions or interactive design and review meetings.

You can also use collaboration tools to accelerate growth by utilising the skills and knowledge of suppliers, partners and customers in an 'extended enterprise'. If you are outsourcing functions like engineering, manufacture or logistics to quickly gain greater scale or reduce costs, you can use collaboration tools over secure networks to control and manage the external operations through performance reviews, scheduled meetings and data sharing. Collaboration with outside organisations offers many important benefits. You gain access to scarce skills not available inside your own organisation or additional resources to tackle larger development projects or reduce lead times. You can take advantage of technologies, components or supplies that enhance your own products or gain opportunities to enter new markets where the partner has an established presence. You may also utilise partners' logistics channels to improve distribution.

The cost of collaboration has discouraged some companies from realising the full benefits. However, cloud-based collaboration solutions can bridge the gaps between offices, companies and geographical locations regardless of technology. That can reduce the cost and complexity of collaboration. Cloud solutions mean that you can utilise collaboration software and networks on a pay-as-you-go basis, rather than investing your own capital in building or buying the right tools.

Customer-Focused Innovation

Customer-focused development is a process designed to ensure that products meet customer expectations from initial concept to the end of their life. Customer-focused development is a form of quality control that ensures product reliability and subsequently improves customer satisfaction and profitability by aligning business processes with your strategic goals and the needs of your customers. The discipline of customer-focused development ensures that you meet customers' expectations at every stage in the development, manufacture and support of a product.

Through-life costing is an important aspect of customer-focused development. Through-life costing refers to the costs a customer incurs in acquiring and using a product. By concentrating on quality, reliability and support, customer-focused development can help to minimise the costs of downtime, repair and replacement customers face when products fail, reducing their through-life costs.

Customer-focused development provides an important coordinating role. The process provides technical coordination to integrate the work of all the different departments that design, develop, manufacture and support the product. The process also brings the customer perspective to all individual product-related tasks. Customer-focused development can reduce your company's product-related costs. At the development stage, the alignment of customer requirements and technical specifications minimises the amount of rework or unsuccessful concepts. By improving product reliability, customer-focused development also helps to reduce your after-sales support costs. Reliable performance leads to fewer support requests, enabling you to reduce your telephone and field support infrastructure.

Customer-focused development helps to reduce the level of risk in new product development by integrating customer requirements with new product strategy. At the initial concept stage, you can focus on ideas that are likely to deliver market success. Successful new products derive from a focus on market requirements, compared to those that result from basic research for its own sake. Measurement is an important element of a customer-focused development strategy. Metrics should cover factors such as: levels of rework in product development; changes in customer satisfaction levels; numbers of product returns or recalls; reduction in service requests.

External Sources

You can source innovation outside your company. There is no shortage of new inventions – a total of 482,871 patent applications were filed with the US Patent & Trademark Office (USPTO) in 2009. The challenge is to find and develop the inventions that will make it as successful new products. Look for inventions and new products that meet consumer needs. Invention for its own sake does not guarantee success as the USPTO figures show. Use market research to identify the strongest trends in consumer needs. Check the websites of patent licensing agents – they list companies looking for inventions in different sectors and that information can highlight gaps in the market.

Contact suppliers, business partners, research organisations and universities to source news ideas or to assess the commercial potential of projects that they are working on. Check patent licensing organisations to identify inventions that might meet consumer needs in your chosen market sectors. You can license existing products or technologies from other organisations to expand your product range, rebranding the products or offering them to consumers as own-label products.

You must first evaluate any invention or suggestions to see if they provide a unique solution to a real problem. Compare the proposal to existing products and assess whether the differential would create a product that could generate revenue and profit. You can locate and work with inventors who have promising ideas but don't have the resources or the funds to turn their inventions into commercial products. To bring a product to market may require design engineering, prototyping, production, marketing and customer service resources. Assess the costs of developing a full product.

The route to market for new inventions takes two distinct directions – licensing your invention to another organisation or marketing it yourself. That decision requires a careful review of risk and reward. Although marketing an invention yourself appears to offer greater control and potentially greater profit, the risks involved in development, production and marketing could be high. Licensing transfers the risk to another organisation with the rewards to you in the form of a one-time payment or continuous royalties on sales. Whatever route you choose, take professional advice at each stage. Work with patent attorneys to register and protect your intellectual property before opening negotiations with any other party. Discuss your invention with patent agents who can advise on the current new product requirements of their manufacturing and marketing clients. Review existing market research sources and consult market research companies on the best way to fill any gaps in the research.

If your invention is an idea or concept, rather than a finished product, carry out market research to assess its potential as a commercial product. Identify customer needs and assess the physical product features your invention would require to meet those needs more effectively than existing solutions. You will need to carry out full development and production costings to determine realistic pricing and volume levels. Identify partners who can bring essential skills that will help you take your invention to market. Universities have innovation groups who carry out basic research or spin out new ideas into commercial products. Many have business incubation units that provide business-related services and facilities to individuals or start-ups that want to develop their inventions. Contact product development companies or marketing agencies specialising in innovations.

Licensing carries a lower risk. Development, production and marketing require specialist business skills as well as considerable financial investment. By licensing an invention to a third party, the inventor is free of the business risk but can still receive an income. The licence can be based on a patent, copyright or just an idea. It's important to work with a company that has an established market presence with the resources and knowledge to market and distribute your product effectively. The better they perform, the more you make in royalties.

Summary

Innovation plays an important role in technology marketing, contributing to the development and improvement of products and customer service standards. Innovation can come from many sources inside and outside your company, but it is important to encourage innovation and support it with collaborative tools that can turn ideas into new products.

11

Make It Easy To Do Business

Improving the customer experience is good for business. Research indicates that businesses would recommend a company that provided an outstanding customer experience. Achieving that level of quality takes a continuous effort to keep improving the way customers interact with your company. Improving customer experience can help to increase sales to existing customers, improve customer satisfaction and improve customer retention levels.

Infrastructure

A customer experience infrastructure provides a framework for managing the customer relationship whenever a customer contacts your company. Gather customer data on website visits, browsing habits and requests to build a profile of their information needs. You can use that data to offer personalised services and recommendations when they visit a website. Give customer-facing staff access to the customer profiles so that they can offer a better quality response. Enhance the profiles by monitoring feedback on social networks to identify important information needs.

You must ensure that customers receive a consistent experience each time they contact your company. Give customers a choice of channels to communicate – telephone, email, fax or web – and provide access to the same knowledge through each channel. Provide a single point of contact in each channel for all product or service enquiries. You can add to the quality of the experience by providing useful knowledge, tips and resources that customers will find valuable. Make the information easily accessible on a website and provide tools to help customers search and retrieve information. Build community facilities on the website so that customers can share information and contribute to the increased value of the site.

Monitor the quality of customer experience and respond to any feedback or concerns about poor experience. Research indicates reported that a high proportion of customers stop dealing with a company because of a bad experience, and many related their bad experience to other people. Industry experience indicates that customer responsiveness delivers excellent customer experience. Companies with strong capabilities and competencies for delivering customer experience excellence are outperforming their competition in terms of customer satisfaction. An effective customer response strategy enables your company to deliver the highest standards of customer care at a time when the customer most needs it. That level of care makes an important contribution to customer satisfaction and long-term loyalty.

Customer Response Strategy

A customer response strategy sets out procedures and quality standards for handling customer incidents and enquiries and identifies the essential infrastructure and skills. The strategy meets a wide range of objectives. It must provide the highest levels of customer support during an incident and minimise inconvenience for the customer. It will ensure that incidents are resolved promptly within agreed timescales and provide customers with quality response and support throughout an incident. It should also ensure that support resources are deployed effectively to maximise customer satisfaction.

Customers can contact a company by telephone, email, fax or web channels. They may want to make an enquiry, order a product, place a service request, report an incident, query an invoice or make a complaint. That means customers could be contacting many different departments and individuals. Putting a single response strategy in place across the company ensures a consistent customer experience every time.

An effective customer response strategy has two main elements – the infrastructure to deliver the service and the personal skills to provide the right level of customer care. The infrastructure might include communications tools to provide a rapid response to customer queries and put the service into operation, a trained support team to deliver customer service and a control centre to coordinate customer response. The skills requirements would include incident management skills to deal with customers who may be in stressful situations and project management skills to coordinate and implement a response. Customer-facing staff also need technical skills to deliver the service and communications skills to coordinate the elements of the response.

Website Experience

Transactions on your website represent an important aspect of customer experience. Personalised pages on a website make it easier for customers searching for product or service information. After entering information to create a personal profile and product preferences, customers are able to view a series of personalised recommendations, together with independent reviews and price plans. A technology known as a prediction platform takes website personalisation a stage further. The platform analyses the behaviour of visitors on a website and collates the analysis with customer profiles and records. The prediction platform then identifies groups of customers who visit a website or contact an organisation frequently. When a frequent caller makes contact, the system predicts the type of assistance they need and offers them a choice of a highly personalised chat service or self-service facility.

If you offer online purchasing, allow pre-registered customers to bypass a conventional shopping cart and checkout by purchasing a product with a single mouse click. The web server recognises the customer and relates their order to stored credit card and shipping information. The primary role of the 'Thank You' page on an e-commerce website is to indicate that the purchase process is complete. This might be important if order processes are prone to problems. Customers want to know that they have finished the process successfully, particularly if they have made an online payment. However, with more reliable order processes, the Thank You page can now take a more proactive role, by indicating to the customer that they can either leave the website or remain and do something else. Experience indicates that many companies assume that after completing the purchase, the customer will leave the site.

The Thank You page should be more than just a straightforward thank you. The customer has just completed a significant process – buying a product, downloading a publication, signing up for a subscription or registering their details. When the customer is in a buying mood, that represents a great opportunity to go further. The Thank You page offers a great opportunity for positive communication. The basic Thank You process takes very little space. There are no other distracting messages and the visitor has provided a great deal of profiling information. With such a captive audience, you can provide a number of alternatives that will benefit both parties. The page should therefore offer a number of suggested routes which might include:

- Continue shopping;
- Return to the home page;
- Find out more about the product;
- Rate the website service;
- Check delivery status.

If your site is designed for lead generation and a visitor has downloaded a whitepaper or other publication after registering information, you now have more knowledge about their level and type of interest. You can use this knowledge to suggest routes to additional content, for example, read a case study about this industry, find out about our solutions for this industry, or contact an industry specialist. This can be a great basis for retaining customers by demonstrating the value of the content on your site. For new customers, it is important to provide careful guidance to the additional content. The more content you can expose them to and the more you can direct them to the right areas, the more likely they are to fall into a pattern of visiting that benefits you and builds long-term relationships.

As well as focusing on exit routes, you can also use other techniques to encourage visitors to stay on a site:

- Suggest related products at a discounted price. If an online shopper decides to buy a particular product, the closer page can offer the product at a special price if the customer buys it together with a related product.
- Suggest linked products. If a customer buys a product that has accessories, ask if they would also like to buy batteries, peripherals or other accessories. Those are items the customer may forget to purchase or might not realise are available.
- Make a limited time offer! A visitor is more likely to take advantage of a special offer if they know there's a time limit.

- Offer to group subsequent orders to reduce delivery costs. It's more efficient to pack more items in each order, so give customers an incentive to purchase more at one time. This may be important if delivery charges represent a significant proportion of a low-value item.

- Offer free delivery for orders over a certain value. This may encourage customers to make additional purchases to reach the required order value.

Follow-up emails are an important part of building a relationship after an online purchase. There should be a series of emails advising the customer of the progress of their order, including:

- Confirmation of their order;

- Advice that their product is in production;

- Notification of despatch.

This keeps the customer informed in a proactive way and reduces the number of delivery enquiry calls your contact centre has to handle. You can also offer the customer a self-service facility – click here to get an update on your order. This gives the customer greater control over the process and also encourages them to continue visiting your site. When a customer has completed a purchase or other transaction, offer them the opportunity to register for other relevant services:

- Set up an account – although a customer may have used a one-off payment method, such as a credit card, offering them an account can encourage repeat purchase, as well as providing valuable profiling information. The account should offer the customer added value such as membership discounts or other privileges, as well as a more convenient method of ordering and payment. The account would also give the customers facilities for reviewing their purchase history, checking delivery information or making returns.

- Offer product notifications – this is particularly important where product changes are frequent and customers can upgrade easily. Customers agree to the notification, providing an opportunity for future communications. In return, they should be offered privileged discounts.

- Offer support services – customers may need a range of services depending on the type of product. These could include installation, telephone support, routine maintenance or repair. This can provide additional revenue and strengthen the customer relationship.

- Offer advice – customers may want advice and help in making the best use of their product. Direct them to a forum or other online community where they can get information, advice and support from other customers as well as experts. They can also sign up for regular newsletters that provide ongoing advice and guidance.

Contact Centre

It's essential to enhance the customer experience across all communication channels and the contact centre is a key area for focus. Advances in technology, such as voice recognition enable organisations to automate a number of processes that contact centre agents handled in the past. Customers answer a series of recorded questions to perform routine tasks, such as billing enquiries, online payments and service requests, without speaking to an agent. As well as improving convenience for customers, contact centre operators believe the facility could reduce call volumes.

While technologies like interactive voice response (IVR) systems, computer-telephony integration (CTI) and call routing have significantly enhanced the customer experience, customers continually demand further improvement. One of the greatest contact centre challenges is responding to fluctuating traffic. Many contact centres today struggle to meet spikes in demand or utilise idle resources when traffic drops. These challenges arise due to unpredictable events, for example, calls to Internet service providers often spike sharply when there is a network problem.

Unfortunately, contact centre staffing levels are generally decided well in advance, making it difficult to respond to unexpected events. On the one hand, allocating contact centre agents conservatively can lead to poor service when traffic spikes. As queue times rise, customers become frustrated, and companies experience high levels of abandoned calls. On the other hand, overscheduling agents raises costs and creates agent dissatisfaction as they sit idle or perform less valuable tasks.

Another challenge is to match different contact centre tasks to individual agent's capabilities. Some contact centre agents are expert at handling inbound calls, while others are equally comfortable handling inbound and outbound calls. Others may prefer to handle only non-voice channels such as email, web chat and instant messaging. Having agents focused on one interaction creates imbalances whenever a particular interaction channel peaks.

To meet those fluctuating challenges, it's important to adjust the volume and mix of available resources to meet changing conditions. At a basic level, agents could shift between activities according to their particular skill set and availability to handle multiple interaction types or manage both inbound and outbound calls. As a result, the pool of resources would instantly expand for one task and contract for another based on analysis of changing contact centre conditions. Another approach is to reallocate staff from other areas of the business or other locations. These resources could be deployed at short notice to help manage high traffic volumes or handle complicated, specialised calls.

Many customers prefer to interact with a contact centre through non-voice channels such as email, text, chat or website. Integrating these channels ensures that customers receive a consistent customer experience across all contact centre media. This is also important for contact centre management as these alternative channels provide cost-effective options to manage traffic and outcomes by shifting the allocation of resources.

Contact centre key performance indicators (KPIs) are metrics that enable your management team to monitor how effectively and efficiently a contact centre is operating. Traditionally, KPIs have focus on efficiency, measuring productivity factors such as average call handling time. However, customer satisfaction is as important as productivity. To put the emphasis on contact centre effectiveness, you should use KPIs to measure other factors such as revenue per call and first call resolution.

Average call handling time remains an important KPI. Contact centre managers have limited resources and need to make the most productive use of their staff. Average call handling time is a useful measure for comparing the performance of operators handling similar types of call. As an example, if a contact centre's main purpose was handling service enquiries, productivity would be important. As the nature of the calls becomes more complex, the average call handling time KPI can create problems. The emphasis on quantity can affect the quality of experience for the customer. Customer experience is particularly important for contact centres handling service requests. Dell told operators to concentrate on solving problems without trying to minimise call times. Although average call time increased, the team increased the number of incidents they resolved on the first call. First call resolution is an important KPI in that type of environment.

The balance between effectiveness and efficiency is equally important in contact centres where operators have responsibility for revenue. If operators are taking orders for products or services, they should take the opportunity to cross-sell additional products or services by spending more time with the customer. The important KPIs in that environment would be revenue per call, customer lifetime value or customer retention rate. If the contact centre is making outbound sales calls to prospects, the important KPIs would be conversion rate and revenue per call.

Contact centre performance influences the quality of customer experience so it is essential that operators provide a consistently high level of service. Contact centre quality is an important differentiator and influences customers' perception of a brand. KPIs should therefore focus on quality service, using metrics such as customer satisfaction levels.

Providing Finance

For many customers, the decision on how to purchase products is as critical as the decision on what is required. According to Cisco Capital, product financing agreements enable customers to benefit from the use of equipment, not the ownership. It gives them the flexibility to manage budgets with predictable payment streams.

A product financing agreement is a method of funding the purchase of a product, such as a major capital item, by means of a loan or other payment options. Offering your customers a product financing agreement can often help to clinch a deal. By working in partnership with a specialist finance company, you can offer customers a range of finance options tailored to their individual budget and cash flow requirements.

Product financing is becoming an increasingly important part of a company's offer. Customers recognise that financing purchases can bring forward business benefits, while preserving capital to grow other areas of their business. Financing provides essential equipment for predictable payment over an agreed term, protecting capital and improving cash flow. That means customers can put equipment to productive use immediately, reducing costs, improving competitive performance, increasing efficiency and delivering other business benefits that can have a direct impact on their bottom line. That makes product financing agreements attractive to customers.

Constant technological developments mean that companies could find existing equipment becoming obsolete. If customers lease equipment, rather than purchase outright, they can benefit from product upgrade options during the term of the agreement, without further capital investment.

An international stock exchange wanted to enhance its competitive position by improving the availability of real-time market and trading data to customers. It also recognised that it would need to refresh its technology on a regular basis to maintain an edge over competitors. A financing deal was structured that enabled the exchange to refresh technology on an ongoing basis while minimising risk and exposure by providing the option to return equipment at the end of the term.

You can also support customers by financing the total cost of equipment. The true cost of equipment purchase goes beyond the cost of the hardware. It includes the cost of installation, training, maintenance and financing – the total cost of operations (TCO).

If you want to offer customers finance agreements, it's important to choose the right finance partner. The finance partner should have a good working knowledge of the target market sector and the potential benefits of the product offer so that it can tailor financial advice in a relevant way. The best financial partners are willing to embrace a comprehensive programme to serve customers and bring additional value to the relationship.

Managing Complaints

As well as helping customers to do business with your company, you should also make it easy for customers to complain. Set up a complaints form or public forum on your website and provide a free telephone number for complaints. Include clear guidelines for making complaints on any product packaging or service documentation. Industry experience indicates that only a small proportion of customers with problems actually complain because they feel they will not get an adequate response. Creating a transparent complaints procedure demonstrates your willingness to deal with problems.

Speed is essential so you should set standards for response times. Acknowledge the complaint immediately and let the customer know when they will receive a detailed response. You should also outline the escalation procedure for any complaints that you cannot resolve to the customers' satisfaction. Providing a speedy response overcomes negative perceptions of your customer service. Explain the reason for the problem to the customer. If it is a product or service issue, tell the customer what action you will be taking to overcome the problem. Offer a refund or substitute product or service if you cannot resolve the problem immediately. Customers who feel you have resolved their complaint fairly will continue to buy from you.

Social networking has changed the nature and impact of customer complaints. Monitor any complaints or negative comments on social networking sites, respond to the comments and try to create dialogue to counter any negative perceptions. Social networking sites make it easy for dissatisfied customers to share their grievances and communicate negative comments to a very large audience.

If you don't help customers to complain, you may have an inaccurate or over-optimistic perception of customer satisfaction. As a result, you may factor incorrect repurchase behaviour into your future sales and marketing forecasts. Understanding the nature and extent of complaints helps you develop strategies to win back dissatisfied customers and increase retention levels.

Customer Retention

Your business risks losing customers through the effects of competitive activity and natural wastage. An average business loses 10 per cent of its customers each year. By maximising customer satisfaction you can retain customers for longer so that they continue to make a contribution to revenue and profitability. This has a further effect on profitability because the cost of acquiring new customers is much higher than the cost of increasing sales to existing customers.

In sectors such as financial services, the top 25 per cent of customers account for the majority of a company's profits. Focusing on customer satisfaction and loyalty is therefore essential to maintaining profitability. You can maximise customer satisfaction by delivering what is valuable to your customers so that they continue to come back to you to buy more of the products and services that meet their needs.

The longer customers remain loyal to you, the more valuable they are to your business. You can use a metric such as lifetime customer value to measure the impact of customer satisfaction on long-term profitability. Lifetime customer value represents the total profits a customer generates while they continue to do business with you.

Satisfied customers have an indirect impact on profitability when they share their opinions with other consumers. Social networking sites play an important part in shaping attitudes towards a company and its products. By maximising customer satisfaction and encouraging customers to post their views on a forum or product review site, you can attract new customers who use the reviews to make their purchasing decisions.

Summary

Managing the customer experience and making it easy to do business makes an important contribution to maximising customer satisfaction and profitability. By maximising customer satisfaction you can increase the opportunity for repeat sales to customers, while reducing the cost of sales and marketing. Customer satisfaction helps to increase customer loyalty, reducing the need to allocate marketing budget to acquire new customers. Satisfied customers may also recommend your products or services to other potential customers, increasing the potential for additional revenue and profit.

12

Manage the Sales Cycle

Companies rarely buy technology products or services on impulse. The decision to buy is the result of a long process of investigation, consideration and review. Different people will evaluate products and companies. To be able to market and sell effectively and to make the best use of your sales and marketing resources, it's important to understand how your customers and prospects buy, and to recognise which stage of the cycle they are currently in. Although your customers and prospects may vary in size, most companies share a common buying cycle. Those common factors can help you determine the nature of your sales and marketing campaign.

Technology Buying Process

The technology buying decision process involves eight distinct stages. At each stage, different decision-makers may be involved, depending on the cost and strategic importance of the purchase. To navigate the buying decision process successfully, you need to provide the right type of information and ensure that your sales representatives are contacting the right decision-makers. You can also strengthen your position by offering customers advice and guidance at each stage – a process known as consultative selling.

The process begins when a company identifies a need for a purchase. It may want to replace an existing item, replenish stocks or buy a new product that is just available on the market. You can also stimulate a need that the company may not be aware of by advising them of issues and challenges that other companies in their industry face. The buying team next works with the requesting department to firm up on the requirement. Your sales team can provide advice and guidance at this stage by offering discussion papers or inviting decision-makers to workshops or seminars on the topic.

When the buying team has agreed requirements, it prepares a detailed specification that sets out quantities, performance and technical requirements for a product. Your sales team can support this stage by advising the buying team on best practice or collaborating with the buying team to develop the specification. Buying teams then use the specification to search for potential suppliers. They may search the Internet to find products or companies that provide a match to their specification, so it is important that your website features keywords that match your customers' product or service needs.

When the buying team has identified potential suppliers, it asks for detailed proposals from the suppliers. The team may issue a formal document known as a Request for Proposal, or it may outline requirements and invite potential suppliers to make a presentation or submit a quotation. If the product or service has a precise specification, the buying team may simply ask for price quotations. If the product is more complex, it may ask for proposals on how a supplier would meet the need.

The buying team evaluates suppliers' proposals against criteria such as price, performance and value for money. As well as evaluating the product, they assess the supplier on factors such as corporate reputation, financial stability, technical reputation and reliability. You can influence decisions at this stage by providing company information, case studies and independent reports that review your company and products. Before the buying team place an order with the chosen supplier, they negotiate price, discount, finance arrangements and payment terms, as well as confirming delivery dates and any other contractual matters. When the order is complete and delivered, the buying team may add a further stage by reviewing the performance of the product and the supplier. This stage may include imposition of penalty charges if the product fails to meet the agreed specification.

The complexity of the buying process places a fundamentally different emphasis on the role of sales and marketing teams. It's no longer enough to base the success of a campaign on the volume of leads. The real measure is the conversion ratio and quality of business wins. Marketing activity needs to go beyond stimulating interest – it has to play a crucial part in all stages of the process until closure. This is a process called joined-up marketing. Joined-up marketing places a far greater emphasis on the whole sales process, ensuring that all parties communicate with consistent messages that adapt to different stages of the buying cycle. The process of managing sales opportunities with joined-up marketing recognises the role and relative importance of different influencers and decision-makers and changes campaign emphasis depending on their views and concerns, objectives and interests.

Individuals make different contributions to the decision-making process and they have different information requirements. Many companies have adopted team purchasing structures to deal with high-value purchases and it is vital that you communicate effectively with every member of the team. Depending on the value and complexity of the purchase, a decision-making team could include:

- Senior executives;
- Purchasing professionals;
- Technical staff;
- Manufacturing managers;
- Service providers;
- Marketing staff;
- Departmental managers.

Constant research is essential to ensure that you are focusing on the right people at the right time. Purchasing situations are fluid and it would be easy to concentrate on the wrong members. To drive the sale towards a conclusion you need to develop a programme that successfully builds relationships before, during and after the sale, qualifying the process all the way through. If the process works properly, the final negotiations are more of a formality than a pitch to secure a sale. The initial contact, the proposal, visits to the website, meetings and discussions, presentations and support material have all contributed to a process where the decision-makers recognise the value of your company over competitive offers.

Just getting the right message to the right person isn't enough. The aim is to ensure that, at every contact point, your communications build understanding among all decision-makers. When the decision-makers are working as a team in your favour, your chances of success are high. Each decision-maker will have their own agenda, but you should be encouraging a collaborative decision-making process that recognises your wider contribution to customer success.

So, when a chief executive asks a technical director for an evaluation of your product, or discusses the potential financial benefits with the finance director, you can be confident that they will be basing their views on the information you have given them. What you are doing is creating a group of advocates within a customer organisation who share the same perception of your company and your products.

Managing Sales Leads

Lead nurturing increases the value of your sales leads and enquiries to generate more revenue. A large proportion of early stage leads that are poorly managed often end up lost, ignored or discarded. Too much time and money is spent on generating leads only to have them neglected. Lead management is the process that consolidates leads and enquiries from all sources, then qualifies, nurtures and converts them into sales-ready leads. The key is to keep in touch, maintaining interest and providing the information decision-makers need. This takes a well-defined process to nurture your early stage leads and build relationships before the buying process. Content is important here, so aim to keep in contact via:

- Email;
- Newsletter;
- Case studies, reference sites and customer reviews on website;
- White papers;
- Conferences, workshops and webinars;
- Telephone.

The aim of lead management is to create an experience that will encourage prospects to request information or ask for a sales visit. By communicating effectively with prospects, obtaining more information about them and making personalised offers, you can increase the rate at which you convert prospects to customers. The starting point is to create a landing page – the first stage in building a relationship with a prospect. Your landing page content must encourage them to take action. This can take various forms: read product information on the page, request a publication, listen to a podcast or watch a video, subscribe to a newsletter or request a call from a sales representative.

Constant communication is an essential element of lead management. An IT security company ran a major lead nurturing campaign to educate prospects and build confidence in the company as a trusted solution provider. The company offered prospects educational documents about IT security that would help them formulate their requirements and select the right solution. After the prospect has taken the appropriate action on the landing page, the company sent an email thanking them for their request and telling them what the next stage in the process will be. If, for example, a prospect requested a technical paper, they sent an email that included a link to the document. A week after the visitor downloaded the document, they sent another email offering other documents or resources that the prospect would find relevant.

The company used the response to the campaign to collect and analyse additional data on prospects, using the information to personalise later communications. The company analysed prospects' information preferences by tracking the action they took on the campaign website. A simple example was when a visitor viewed technical pages on the website, rather than business-related pages. They personalised future communications by sending or offering technical content. The campaign provided prospects with content that moved them from initial education to specification of their IT security needs and finally to requesting detailed product information. Once the prospect had requested product information, the sales team made contact, recognising a hot prospect. The sales team also had a detailed understanding of the way prospects had moved through the campaign, giving them clear guidelines for the next stage in the sales process.

A campaign like that requires careful management. You need to create a library of content such as briefing papers, newsletters, videos and other material that visitors can download from the landing page. The person responsible for the lead management campaign should develop a workflow that sets out the sequence of events following the prospects' initial click. An example of a simple workflow is: visitor clicks on ad; visitor reaches landing page; visitor is offered brochure; if visitor does not download brochure, visitor receives a follow-up email with alternative offer; if visitor downloads brochure, visitor receives follow-up email offering further resources.

Integrating Sales and Marketing

As that example illustrates, bringing sales and marketing closer together is a means of accelerating the sales cycle and delivering exceptional results. The basis of the approach is to gather and share better customer and market intelligence through research and feedback and use that data to win more customers. The improved understanding of customer issues and challenges makes it easier to create relevant, targeted communications with prospects throughout the sales cycle.

It improves the results from sales and marketing activities by prioritising opportunities, maintaining the momentum of customer acquisition and speeding up the whole sales process. It also makes sales and marketing activities more accountable by measuring the results of campaigns against target revenues and numbers of closed customers.

To maintain campaign momentum, it's essential to communicate regularly and effectively with all prospects throughout the process. That requires ongoing research and collaboration between sales and marketing to gain a better understanding of the concerns and information needs of different decision-makers. By working together and understanding the needs of the sales process, marketing can drive a bigger impact on sales by matching marketing communications to the time that is right for the sales process.

Changing roles, responsibilities and priorities within a decision-making group can impact on the success of a sales and marketing programme, so it is essential to maintain up-to-date information on key prospects. By continually gathering intelligence, sales and marketing teams can improve their understanding of prospects, their objections and any other barriers to a sale. The information will also help to build clearer sales profiles as a basis for focusing sales effort on the most appropriate prospects. This requires an intelligence plan that gathers and builds essential insight throughout the sales and marketing process, covering key prospect data such as:

- Prospects' challenges;

- Prospects' objectives and goals;
- Levels of prospect awareness;
- Competitor information.

The key to success is using intelligence to fine-tune communications. This can ultimately lead to a capability to personalise messages for each key prospect. By obtaining regular feedback, organisations can also identify a prospect's propensity to buy at different stages of the sales process. Communications can then be modified, with messaging refocused to reflect individual interests and concerns. This, in turn, will provide essential support to the direct sales effort and help to convert more prospects to customers.

Marketing Waves

When your company is planning marketing strategy, it's important to relate individual sales and marketing activities to overall strategies. A marketing wave is an integrated go-to-market plan that links sales strategies with lead generation, advertising and other marketing activities. In some cases, marketing waves can cross product and organisational boundaries with local campaign variations to ensure overall success. A marketing wave describes the integrated strategy you develop to connect with your customers and the processes you use to manage customer interactions from initial contact through to order fulfilment. The process covers:

- Commercial strategy, objectives and goals;
- Commercial execution across marketing, sales, pricing and customer insight;
- Go-to-market capabilities including organisation and process.

Marketing waves are designed to achieve measurable results and contribute to your company's strategic goals for revenue growth. They bring together all the assets that are essential to achieving those objectives in a single integrated plan.

A marketing wave enables your company to deliver its unique value proposition to the target market. The marketing wave identifies who your company will target, what it will offer customers, and how that will happen. A marketing wave takes into account the following factors:

- Market demand – what is driving or inhibiting market demand for your company's products?
- Customers – what are the customer needs and challenges and what are their key decision-making factors?
- Competition – who are your main competitors and how well are they performing?
- Differentiation – what differentiators do you have? Are they sustainable and difficult to replicate by competitors?
- Solutions – what solutions best meet market demand and customer needs?
- Channels – what channel structure enables you to reach your target market cost-effectively?

A marketing wave brings together all the commercial functions – sales, marketing, brand management, pricing and customer insight – into an integrated strategy that is aligned with business objectives and focused on profit.

This approach provides a number of important benefits:

- Ensures that your go-to-market strategy fits the current needs of your business objectives, channels and customers.
- Improves alignment and integration across the various commercial functions.
- Supports cost-reduction efforts without harming the core business.
- Brings together the right processes, people and infrastructure to support an effective go-to-market strategy.

- Improves existing marketing and sales capabilities and helps you increase competitive advantage.

- Positions your company for sustained marketing improvement.

When sales and marketing campaigns are managed on an ad hoc basis, responsibilities are poorly defined, and the result is tactical go-to-market plans and constant, disruptive requests for sales support. With integrated marketing waves, your sales and marketing teams can optimise their resources and deliver more effective results by collaborating throughout the sales cycle and focusing on the same business and strategic goals.

Marketing waves can help companies move from a product-led sales and marketing operation to one that is focused on customers' business, technical and financial needs. Marketing waves start by building an understanding of the issues and challenges facing all members of the decision-making team and developing targeted communications and marketing campaigns that address those needs and move prospects towards a purchase decision. Marketing waves bring together disconnected sales and marketing activities, creating a structured campaign that contributes to your overall corporate goal. They bring together sales and marketing assets in a plan that is designed to achieve measurable results. This enables you to build a strong business case for the campaign, set the right budget and justify the marketing investment. Marketing waves contribute to your company's strategic goals for revenue growth, positioning awareness and customer retention by focusing on the campaigns that have the highest probability of contributing to measurable improvements in both short and long-term goals.

By integrating marketing waves, you can help both direct and indirect sales channels to succeed without providing ad hoc support for individual deals. Marketing waves incorporate elements to move your prospects through every stage of the sales process from lead generation to completion. That can improve sales force productivity as well as making more efficient use of your marketing resources. In most companies, the individual sales and marketing programmes are likely to be in place. Using a marketing waves approach brings those elements together in an integrated strategy that delivers improved results.

Marketing waves should have a specific purpose within your overall business and marketing strategy. In the example that follows, the company's overall objective is to strengthen relationships with its customers by marketing a series of high-value services that will increase customer dependency. The company develops a series of marketing waves that move the customer through the initial introduction of its services to a situation where the company is managing all of its customer's strategic services. Their aim is to demonstrate how attractive their service value proposition is to the customer. The waves consist of a planned set of offerings that begin with an entry offering followed by other waves that create demand for other service offerings. Together, the waves provide a framework for building a planned, structured approach that communicates the complete value proposition offered by your company.

To launch the services, the company operated four specific marketing waves:

- Entry offer – to create interest and awareness of the service offer;
- Assessment – to encourage customers to request an assessment to show how the service can deliver benefits;
- Pilot project – to demonstrate the results of the service;
- Managed service offer – to encourage customers to take the full service package.

To stimulate initial interest, the company focused its efforts on business decision-makers, communicating potential benefits in business terms, not technology terms. The communications highlighted a key business problem and encouraged the prospect to take action by requesting whitepapers, subscribing to executive briefings, seminars and other information sources. To add value and credibility to the entry offer, the company used an independent authority to deliver the key messages. The aim of this wave was to demonstrate that the company understood the key challenges facing its customers and had the capability to deliver a solution to those challenges.

At this stage, the company wanted to secure appointments to carry out assessments with those prospects that had requested initial information. As well as carrying out the assessment, the company used case studies and customer referral sites to provide proof that their services could deliver business and operational benefit. The company worked with prospects to put together a business case for the service, using input from the case studies. In this wave, the company focused on organising a pilot project and helping prospects to assess the results from the project. The pilot project helped the company to lay the foundation for introducing an additional range of services. By demonstrating value in the pilot project, the company was able to upsell other services and increase its influence with the customer.

The final wave was designed to build long-term dependent relationships with customers by offering them managed services – a solution where the service provider takes responsibility for running the customer's service-related activities and acts as a strategic partner.

Supporting Sales

The importance of integration between sales and marketing means that technology sales managers require a range of additional skills to improve their own performance and the performance of their team. They must be able to sell at many different levels in the customer's organisation, understanding the concerns and challenges each member of the buying team faces. They must also develop consultative selling and key account management skills to deal with complex sales. Detailed product knowledge is essential to every member of the sales team, so it's important for sales managers to work with the product and marketing teams to develop training sessions on new products in the company portfolio or emerging technologies that will influence future product developments. By getting involved in product planning meetings, they can improve skills in developing complete customer solutions that include products, services and training.

The purchasing team for information technology products includes members from technical, business and financial disciplines so sales managers need to develop skills and use tools that enable them to communicate with different decision-makers in their own language. Your marketing team can support this by preparing presentations on the business case, cost-benefit analysis, Return on Investment and other metrics that could influence the buying decision. Information like that demonstrate value to prospects, so it's important to build a profile of prospects and the challenges they face as a basis for designing a comprehensive market offering that represents a convincing technical, business and financial solution.

Sales Automation

The length of time and complexity of information technology sales makes account management skills as important as face-to-face selling skills. Success in account requires managers to gather information, analyse the competition, develop strategies and execute strategic account plans. However, industry experience indicates that sales representatives spend a significant proportion of their time on non-sales activities. By automating routine administrative tasks and putting relevant information on the desktop, you can free your account team to spend more time on productive work.

Sales automation tools allow your sales representatives to quickly identify promising leads, and then coordinate the account teams' efforts to close the deal. They help you lower your cost of sale, respond faster to customer enquiries and share information across your entire company. For example, sales teams can use automation tools to gather up-to-the-minute customer information over the Internet before calling on potential customers. They can then use this data to customise product offers or provide quotes in real time during the sales calls. They can instantly forward orders over the Internet to head office. This speeds delivery and accounting processes, and increases sales teams' ability to respond to customer needs.

Sales automation tools support key aspects of the selling process, including marketing, contact management, account management, prospecting, order fulfilment and customer service. By implementing Internet-based sales automation tools you can:

- Streamline sales processes;
- Boost sales force knowledge;
- Enhance collaborative selling;
- Improve customer relationships;
- Reduce quote times;
- Increase sales force morale.

Online product catalogues can be updated the moment a new product or service is available. Sales representatives can access product specifications information at their fingertips. These electronic catalogues can also be set up to alert sales representatives to complementary products and promotions, immediately creating other selling opportunities. An online sales library can provide staff access to updated price lists, presentations and brochures created by your marketing department. This information can also be emailed to customers and prospects. The system can also incorporate a knowledge base which has information about product issues and modifications.

The system can help to improve sales force performance through the provision of comprehensive sales support information. It can provide the sales force with a wide range of information that builds a complete picture of customers and the way they respond to sales and marketing initiatives. The system can be used to provide the following information:

- Customer profile and contact information;
- Customer purchasing history;
- Any known problems;
- Individual sales or direct marketing initiatives and the response;
- Wider communications programmes or promotional activity within the customers' sector;
- Competitive activity on the account;
- Profitability of the account.

This kind of information improves productivity by focusing the sales and marketing teams on the most important customers and prospects. All users of your sales automation system share access to a single data source, which supports better teamwork between sales, marketing and customer support. The system capabilities can be extended to include sales teams from your distribution partners.

Sales automation systems can be configured to communicate information to your customers based on their specific needs and keep existing customers abreast of product updates. They also support online customer service, including automated help and access to information 24 hours a day, 7 days a week. Your sales representatives can use automation tools to determine customer requirements and immediately provide a complete proposal after a single meeting, cutting days or even weeks from the selling cycle. Some systems incorporate accounting links that enable the sales team to service customers faster by providing access to up-to-date information on customers' credit limits and balances, recent quotations and invoices. The system can also create estimates, orders and up-to-date price lists automatically.

Summary

The sales cycle for technology products is long and complex, and success depends on close integration of sales and marketing activities. Marketing provides the information that different members of the decision-making team require at each stage of the purchase cycle, while feedback from the sales force enables marketing to fine-tune communications and help to accelerate the sales process.

13

Enhance Marketing Management

Marketing is a critical activity for technology companies, so it's essential that marketing is managed at a senior level. The appointment of a marketing director or chief marketing officer (CMO) at board level plays an important role in the top management team. As well as coordinating all your company's marketing activities, the CMO brings the perspective of the customer to the important strategic decisions made at board level. Decisions about investments, acquisitions, recruitment, organisation structure and product strategy must reflect the needs of your customers. Appointing a CMO to the top management team ensures the right level of customer focus.

Strategic Management

CMOs make an important contribution to strategy development. They bring an understanding of markets and customers that can help your top management team make decisions about which markets or segments to enter, which products to develop and which companies to approach as partners. This insight is important for making complex decisions about competitive markets where change occurs very quickly. Successful companies have a customer focus throughout the organisation. By appointing a CMO to your top management team, you demonstrate the importance of customer focus and you can lead customer satisfaction initiatives from the top. CMOs recognise business opportunities and can help your team allocate and prioritise resources for research and development and marketing communications.

Your top management team faces complex decisions about product strategy. Your team needs to understand where various products are in the product life cycle. A CMO can help decide the level of marketing resources you need for products in the launch, growth or maturity phases. They can also advise on the best strategy for dealing with products in decline – should you make improvements and relaunch the product or should you invest resources in developing a new product.

The presence of a CMO on your top management team can build a better understanding of the contribution of marketing to revenue and profit. When marketing is accountable for bottom line responsibility, other members of the top management team will recognise its importance and allocate resources to ensure its success.

Adding Resources

The complexity of technology marketing and the wide range of marketing tasks mean that your company will require a broad range of skills. However, there will be situations when you need skills and resources that are not available from your team. Programmes such as exhibitions or events can impose considerable pressure on the existing team, distracting them from their main priorities, and sometimes leading to poor results. One solution would be retraining, but that takes time and deadlines could be tight. You could also consider recruiting full-time staff with specialist skills to manage projects such as roadshows or website development. But, if these projects are only short-term, you may find that the specialists are underused and this could prove an expensive solution because of the fixed costs of recruiting, training and employing permanent staff.

Technology companies can utilise a recruitment process called resourcing – hiring specialists for the period of a specific project – which provides a flexible alternative to retraining or permanent recruitment. Resourcing is ideal for acquiring specialist skills on a short- or mid-term basis for projects such as events, conferences, product launches or other tactical campaigns. It also provides a cost-effective solution to the inevitable peaks and troughs in marketing activity.

Resourcing differs from contracting. If you take on contract staff, you are normally required to hire them for a fixed period – three months, six months or a year. Resourcing gives you the option of taking on the skills you need for short or long periods, depending on the project duration. The other major difference is that the resourcing specialists work directly for a recruitment agency, but are based in your marketing department as an integral part of your team. Resourcing is focused on results, with the emphasis on delivery and this gives the resourcing specialists a greater sense of involvement in the work and success of your team. However, resourcing staff are outside the politics that go with being on permanent staff. They are able to focus their efforts on getting the job done rather than getting embroiled in office politics or team bureaucracy.

Resourcing adds to the capability and experience within a team and complements it. Your marketing department can use resourcing to acquire specific marketing expertise without adding headcount. The types of people who go into resourcing have the market insight to bring a depth of understanding of the decision-making processes and the key challenges you face. They can work effectively with other members of your team and identify the marketing programmes that will enable you to meet your goals. To enhance your business development programmes and achieve an effective return on marketing investment, they bring strategic vision and an ability to work effectively with your sales, marketing and product development teams. They apply acquired industry knowledge and experience of what works well, and what doesn't, from previous employment. That allows them to hit the ground running and help you achieve results in essential areas.

By taking on resourcing staff for those special projects, your department can acquire skills, services and flexibility without the hassle of overheads, sick pay, redundancy payouts and other recurring employment costs. The resourcing model removes 'people costs' such as pension and private health contributions, as well as training, recruitment, human resources management, payroll and social costs from the equation. The savings can be significant, particularly when typical recruitment agency fees of 20 per cent are taken into account.

While some specialists are in constant demand because of their outstanding skills, and earn substantially more than internal staff, the overall cost of hiring in project-based resourcing staff is likely to be competitive with internal team costs. Another advantage is that the cost of resourcing staff can be charged to a separate services budget, removing it from fixed overheads. That improves financial flexibility, particularly when there is pressure to reduce departmental budgets.

A telecommunications company was aiming to cut its costs by reducing headcount across all departments. The marketing department faced the prospect of losing two experienced executives who would be essential in meeting their targets over the coming year. Their solution was to make the executives redundant, which met their internal requirements. The executives then joined a resourcing agency who hired them back to the company with payment from a marketing services budget. The company was able to show reduced overheads, but was able to maintain the momentum of its essential marketing programmes.

If you need to create a more responsive marketing environment to adapt to changing market conditions, accelerate rollout of strategic programmes or improve return on marketing investment, resourcing offers a highly efficient, cost-effective alternative model that offers important business benefits while reducing costs, improving response to market opportunities, accelerating programme deployment and maintaining marketing momentum. Companies who have implemented resourcing report a number of important benefits, including:

- Reduction in programme deployment times;
- Quicker response to market opportunities;
- Savings on consolidated marketing service costs;
- More dynamic, flexible marketing team;
- More effective solution than recruitment or stretching existing resources;
- Enhanced team performance by providing diverse or specialist skills on a short- or mid-term basis;
- Cost-effective solution to peaks and troughs in marketing activity.

A company with its own internal advertising department found that it was unable to cope with the changing pattern of communications. The team was experienced in press and television, but didn't have the skills to handle online campaigns, e-marketing or social media. Pressure on budgets and a demand for better return on marketing investment meant that they had to deliver results and demonstrate that they were adding value across all media. Resourcing provided the back-up they needed for their long-term assignment of developing an online presence and for a series of critical email campaigns.

Events contribute to the success of product and service launches, partner activities and product communications. However, they can also represent a drain on budgets and resources, so they must be highly focused with tangible and demonstrable benefits. A components company had committed to run a series of local dealership events which were proving to be time limited, but resource intensive. The marketing department was struggling to give the events the time and commitment they need. A resourcing agency provided them with an event specialist who delivered the right mix of organisational, logistics and planning skills to manage the programme over a nine-month period. The events helped to drive traffic to the dealerships and contributed to revenue, without tying up the marketing department's own resources.

Focusing on a specific sector or regional market significantly improves your marketing precision and return on investment. It's vital to fully understand the current issues and challenges for individual niche markets, whether you segment them by horizontal or vertical sector, size of business or region. The tight focus means you don't just communicate the relevant benefits and value points of your products, you also deliver them in a way that recognises the concerns of key decision-makers in each sector. However, as a network security company realised, it can be difficult to find that detailed sector knowledge in a team with general marketing skills. The company wanted to establish leadership in a number of different sectors, with a particular emphasis on the public sector. Through a resourcing agency, they were able to bring in a marketing executive with experience in that sector. Over a six-month period, the executive built up effective liaison with many different internal departments, transferring his own sector knowledge. With valuable client contacts in the sector, the executive was also able to add value to the company's lead generation campaigns.

Increasingly, companies are looking towards alliance relationships as the route to strategic growth and enhanced competitiveness. However, many businesses do not have the resources to establish and manage a strategic alliance. A biotechnology company that wanted to move into new markets recognised that it would need to work with partners to source the technology and services needed to succeed. It used resourcing to bring in a professional with the skills and experience to manage the alliances and partner development activities. The executive set up an alliance infrastructure and took responsibility for the initial execution and management of co-market development activities. As a result, the company was able to attract partners who were leaders in their field and who recognised that the alliance would prove mutually beneficial.

Consultancy Support

While resourcing staff can provide you with the additional project expertise you might need from time to time, you might also need high-level expertise to tackle strategic marketing challenges or the problems of change.

Marketing consultants can support existing marketing programmes, accelerate progress or create new business opportunities by providing expert advice, guidance, counselling, coaching or project management. Consultants bring the experience and capability to turn objectives into plans, action and results. They can bring clarity to difficult decisions through their first-hand experience of the challenges facing growing businesses, especially the need to increase profitability through improved alignment of marketing, sales, brand and product development activities.

Consultants can also offer advanced skills in applying strategic marketing techniques to successfully steer your business into new territory. They can act as a catalyst to help your company rethink its marketing strategy to meet new opportunities or keep pace with market changes. They can also help your business grow and become more profitable by pursuing new business initiatives to boost growth, perhaps entering new markets, introducing new products or establishing innovative services. A consultant's independence from internal issues enables them to offer a more objective view. It can also provide a level of authority that can cut through inertia and stimulate productive action. This external perspective can be particularly useful when tackling topics such as branding and messaging, where internal staff may be too close to provide a customer-focused perspective.

Good consultants also transfer knowledge, encouraging your internal team to grow in experience and confidence. They can provide a rich source of information and knowledge of marketing strategy and programmes that improve internal skills and understanding. Many consultants publish original thought leadership papers, practical advice and references to useful marketing resources, as well as regular news and updates on marketing through newsletters. Consultants' websites often represent a valuable repository of insights and techniques, based on their experience of running marketing programmes and delivering consultancy.

Hiring consultants or specialists on a resourcing basis can provide your team with the additional skills needed to tackle time-consuming or difficult projects. Resourcing has proved to be a cost-effective solution that can deliver effective results and save money. It meets the needs of many different types of project and gives you the opportunity to move into new markets or focus on projects or activities that have previously been neglected. This approach works well for managing individual projects, but, if your company wants to take on resources on a larger scale, you could consider outsourcing to a specialist marketing firm.

Outsourcing Marketing

Outsourcing saves you money by cutting out the costs of an internal marketing department. You can spend the money on marketing campaigns, rather than employees. Using a specialist marketing firm can improve the results of your campaigns, increasing revenue. Outsourcing also lets you take advantage of marketing opportunities immediately, creating further revenue opportunities.

Outsourcing eliminates the salaries, benefits and other costs of recruiting, training and retaining experienced marketing staff. You gain access to highly qualified marketing professionals in outsourcing companies by diverting your marketing budget from overheads to campaigns and programmes that deliver results. You can balance internal and external resources by retaining key staff with strategic marketing and project management skills to coordinate the work of outsourced suppliers. You will have access to specialist skills that you cannot afford to hire permanently and take advantage of short-term revenue opportunities by outsourcing marketing. Outsourcing firms can take projects and run with them. Planning and implementing programmes internally would take time and reduce the window of opportunity.

Outsourcing can drive a better return on marketing investment. Specialist marketing firms are likely to deliver better results than employees with general marketing skills. Marketing teams are under pressure from boards to justify expenditure and demonstrate value, so you should measure campaign and programme results to identify the return and arrange to pay the outsourcing firm by results. Payment by results is becoming a popular arrangement between clients and agencies, particularly for online campaigns.

If you don't want to use external marketing resources, you can overcome a number of marketing challenges by automating routine marketing tasks, providing the data you need to target customers more precisely, opening new channels of communication and simplifying the distribution of marketing material. Marketing software enables you to automate time-consuming tasks such as managing email campaigns. The software distributes emails automatically at predetermined intervals and provides reports that indicate the number of people who open emails or click on links within the email. Programmes incorporate features that enable you to respond to different types of user behaviour. When a prospect clicks a link, the software sends an auto-reply, together with the requested content. You can also set rules that enable the software to send different responses to different types of user, for example, a business briefing to a chief executive, or a technical paper to a chief information officer.

Marketing Automation

Technology can also help you increase marketing precision. The more information you have on customers and prospects, the more precisely you can target communications and offers. Data storage techniques and data networks allow you to store, manage and access vast amounts of market and individual prospect data for campaign research and planning. You can use your internal network to gather data from every type of customer interaction – email communication, telephone orders, service requests and website visits – to provide a comprehensive picture of the customer relationship. Technology has also opened new channels of communication with your customers and is helping to build dialogue. Traditional channels such as advertising or direct marketing represent one-way communication, with no opportunity for feedback or dialogue. By using social media such as website forums or product review pages, you can find out what your customers are saying about your product and your company, and you can engage in dialogue by responding to comments.

Distribution and control of your marketing material can prove time-consuming. If you provide marketing material, such as brochures or advertisement and email templates to your distributors, you can provide access via a secure portal on your website. Distributors log in and can download different types of material, depending on their status. Authorised distributors, for example, would be able to download all material; an independent distributor might only be able to access product information. This gives you greater control over the way other parties use your marketing material.

Cloud and Marketing

Cloud computing can help you improve marketing performance even further. IT resources are increasingly important to the success of marketing campaigns, but you may only need the resources for a short period. Cloud computing offers a pay-as-you-go approach to getting the short-term IT resources you need to support a marketing campaign. You simply hire the resources you need and you only pay for them as long as you need them.

A mobile device manufacturer wanted to set up a pilot regional distribution network to test the company's expansion plans into international markets. By delivering cloud-based online ordering and customer relationship management resources to a pilot group of retailers via an existing telecommunications network, the manufacturer was able to test the market before committing permanent resources to the programme. The project was planned as a self-funding exercise; if the test market succeeds, the incremental revenue covers the cost of hiring the cloud services. If it doesn't work out, the company would simply shut down the cloud service at the end of the project with no further cost and no loss of scarce investment funds.

That flexibility is ideal if you have a regular programme of new product development. Companies that emphasise new product development explore new business opportunities followed by rapid growth into new markets when research highlights a profitable opportunity. But, since many new ideas don't make it to market, they need to try a lot of ideas to find the winners, then find ways to bring them to market quickly and cheaply before committing full-scale resources. Waiting for the deployment of permanent IT resources slows down the process and that could be risky if time to market is critical. With cloud computing, you pay a fee and the additional IT services are available immediately to support any campaign where you need extra IT resources quickly.

If you run a successful campaign and find your website is overwhelmed with responses, you could find that your potential customers are frustrated and abandon the site while you wait for the IT department to complete an upgrade. That would be a waste of campaign effort and budget. You can pull in the additional capacity via the cloud and eliminate the delay. At the end of the campaign, you switch off the cloud-based resources and there are no additional IT costs. That facility would also prove vital if you face a product recall. IT problems could create a public relations disaster if customers can't get through. Cloud gives you the additional IT resources you want immediately. The speed of response could turn a potential public relations problem into a boost for customer satisfaction.

For many companies, collaboration over the cloud is playing an increasingly important element in marketing strategy. Partnership with suppliers and customers has become an integral part of the product development process. The problem is that as the ecosystem gets bigger, the task of trying to collaborate with partners who might be located anywhere in the world over a variety of different networks, devices and communications standards gets more and more complex. The cost and complexity of collaboration has discouraged some companies from realising the full benefits. Moving collaboration to the cloud makes it possible to solve that challenge by bridging the gaps between offices, companies and geographical locations, regardless of technology. Easier collaboration can have a big impact on product development timescales and effectiveness.

These operational advantages are just part of the way cloud is helping to transform marketing performance. Add in all the cloud-based marketing automation tools that are already mainstream and it's clear that cloud computing is no longer a technology issue. It's an important competitive differentiator and it looks set to bring many benefits to the marketing department.

Managing Marketing Technology

As technology becomes more important to the marketing department, it's important to have people in the team who understand the contribution it can make. Marketing consultant Scott Brinker introduced the term chief marketing technologist in an article, '8 Things Every Marketing Technologist Should Know'. Brinker believes that the combination of skills and resources from marketing, information technology, software development and digital media can help companies improve marketing performance. To coordinate those elements, he recommends appointing a chief marketing technologist with experience in all the fields. Small companies, who may not have the resources to recruit a specialist, can apply Brinker's findings and gain significant benefits.

Scott Brinker suggests that the person with the role of chief marketing technologist should be responsible for coordination or liaison with specialists in the following areas: data and analytics, marketing software applications, digital advertising, social networks and mobile marketing, content marketing, web browser technology, software programming and IT operations in relation to cloud computing.

Data and analytics are central to the role of a chief marketing technologist. You can apply the principles by ensuring that you gather data on customers and prospects from all your marketing channels, manage the data and use the information to make your marketing communications more precise. You can use the data to improve the quality of content marketing, another of the eight elements by developing and delivering content that is relevant to the information needs you have discovered through data analysis.

A detailed knowledge of marketing software can help your company automate routine marketing tasks such as email distribution and campaign management, customer relationship management, managing sales leads or managing events. Automation can help you implement more marketing campaigns and manage them more effectively, making better use of your budgets and resources. Appointing a team member with knowledge of software programming will help your marketing team communicate more effectively with developers so that they can build or customise marketing software for your specific needs.

Brinker emphasises the importance of digital media as communication channels. Using online advertising, social networks and mobile marketing tools will help you develop a communication strategy that reaches customers and prospects cost effectively, compared with traditional channels such as press advertising or direct marketing. Key digital channels include your website, online advertising, social networking, online communities and forums, email marketing and mobile marketing.

Summary

Managing technology marketing requires wide-ranging strategic and tactical skills. Representation at senior level is important to build customer focus throughout the company and demonstrate accountability. Services such as resourcing, consultancy and outsourcing can provide you with specialist campaign support and help you improve performance while marketing automation systems and technologies such as cloud can improve your productivity and give you more time to concentrate on strategic tasks.

Bibliography

Battelle, John. *The Search: How Google and Its Rivals Rewrote the Rules of Business and Transformed Our Culture*. Nicholas Brealey Publishing 2006.

Brandt, Richard L. *One Click: Jeff Bezos and the Rise of Amazon.com*. Viking 2011.

Brinker, Scott. *8 Things Every Marketing Technologist Should Know*. http://www.chiefmartec.com/2011/01/8-things-every-marketing-technologist-should-know.html

Butje, Mark. *Product Marketing for Technology Companies*. Butterworth-Heinemann 2005.

Corsi, Patrick. *The Marketing of Technology Intensive Products and Services: Driving Innovations for Non-marketers*. Wiley-Blackwell 2009.

Cusumano, Michael A. and Richard W. Selby. *Microsoft Secrets*. HarperCollins 1996.

Davidow, William H. *Marketing High Technology*. Macmillan 1986.

Davis, Stan and Christopher Meyer. *Blur: The Speed of Change in the Connected Economy*. Capstone 1998.

Gates, Bill. *Business @ the Speed of Thought*. Warner Books 1999.

Green, Emily Nagle. *Anywhere: How Global Connectivity is Revolutionizing the Way We Do Business*. McGraw-Hill 2010.

Keyes, Jessica. *Marketing IT Products and Services*. CRC Press 2009.

Malone, Michael S. *Infinite Loop: How Apple, the World's Most Insanely Great Computer Company, Went Insane*. Aurum Press 1999.

McGrath, Michael E. *Product Strategy for High Technology Companies*. McGraw-Hill Professional 2000.

McKenna, Regis. *Relationship Marketing: Successful Strategies for the Age of the Customer*. Perseus Books 1991.

Mohr, Jakki J., Sanjit Sengupta and Stanley Slater. *Marketing of High-Technology Products and Innovations*. Pearson Education 2009.

Moore, Geoffrey A. *Crossing the Chasm: Marketing and Selling Technology Products to Mainstream Customers*. Capstone 1998.

Papows, Jeff. *Enterprise.com: Market Leadership in the Information Age*. Nicholas Brealey Publishing 1999.

Peppers, Don and Martha Rogers. *Enterprise One to One: Tools for Competing in the Interactive Age*. Doubleday 1997.

Ryans, Adrian, Roger More, Donald Barclay and Terry Deutscher. *Winning Market Leadership: Strategic Market Planning for Technology-Driven Businesses.* John Wiley & Sons 2000.

Seybold, Patricia B. *Customers.com: How to Create a Profitable Business Strategy for the Internet.* Century Business 1998.

Shapiro, Carl and Hal R. Varian. *Information Rules: A Strategic Guide to the Network Economy.* Harvard Business School Press 1999.

Sidhu, Inder. *Doing Both: How Cisco Captures Today's Profit and Drives Tomorrow's Growth.* FT Press 2010.

Stross, Randall E. *The Microsoft Way.* Addison-Wesley 1996.

Temporal, Paul and K.C. Lee. *Hi-Tech Hi-Touch Branding: Creating Brand Power in the Age of Technology.* John Wiley & Sons 2000.

Wiersema, Fred. *Customer Intimacy.* HarperCollins 1996.

Index